D0860731

UNIVERSITY OF WINNIPEG
LIBRARY
DISCARDED
Winnipeg, Manitoba R3B 2E9

plants
A Scanning Electron Microscope Survey

Pollen Grain of Storksbill *(Erodium moschatum* (L.) L'Hérit *)*

See Plate 150

QK
641
.T738

plants
A Scanning Electron Microscope Survey

J. H. Troughton
Research Scientist, Plant Physics Section, Physics and Engineering Laboratory, New Zealand, Department of Scientific and Industrial Research

F. B. Sampson
Senior Lecturer in Botany, Victoria University of Wellington

John Wiley & Sons Australasia Pty Ltd
SYDNEY New York London Toronto

Copyright © 1973 by John Wiley & Sons Australasia Pty Ltd

All rights reserved

No part of this book may be reproduced by any means, nor transmitted, nor translated into a machine language without the written permission of the publisher

ISBN and National Library of Australia card numbers:

Cloth 0 471 89115-0

Paper 0 471 89116-9

Library of Congress Catalog Card Number: 73-968

Registered at the General Post Office, Sydney, for transmission through the post as a book

Printed at The Griffin Press, Adelaide, South Australia

Contents

List of Plates

PHYLUM BRYOPHYTA

CLASS HEPATICAE

CLASS MUSCI

PHYLUM TRACHEOPHYTA

SUBPHYLUM PSILOPHYTINA

SUBPHYLUM LYCOPHYTINA

SUBPHYLUM SPHENOPHYTINA

SUBPHYLUM PTEROPHYTINA

CLASS FILICINEAE

Preface

The development of the Scanning Electron Microscope (SEM) has permitted three-dimensional aspects of plant structure to be illustrated at high magnifications, in a manner not previously possible. The plant photos in this book have been chosen to supplement illustrations in introductory botanical and biological texts used in high schools and universities.

Some plants are more amenable to SEM investigation than others and therefore certain groups have received more attention than others. Pollen grains are particularly suited to SEM examination. As pollen morphology receives little treatment in introductory textbooks, we have dealt with this topic in some detail.

The text is mainly confined to an explanation of the illustrations and, in some instances, to an account of development leading up to a stage illustrated. In addition to information which we have obtained in the preparation of this book, we have made reference to recent research articles, to ensure that the text is up to date.

The classification system adopted is, with a minor amendment, that used in Raven and Curtis' *Biology of Plants*.

The co-operation and assistance of numerous persons is gratefully acknowledged, especially staff of the Physics and Engineering Laboratory, New Zealand Department of Scientific and Industrial Research, through Dr M. C. Probine, and of the Botany Department, Victoria University of Wellington, New Zealand, through Professors H. D. Gordon and J. K. Heyes.

We record our sincere appreciation of technical assistance from Mrs K. A. Card and Mrs L. A. Donaldson and the preparation of prints for publication by Mr H. B. Foster. The work, including the preparation of material by freeze-drying (Zwanneveld, 1973), used the facilities of the Electron Microscope Unit, Physics and Engineering Laboratory. Plant samples were obtained from numerous persons and their co-operation is gratefully acknowledged. We thank Professors Gordon and Heyes and Dr J. W. Dawson for reviewing the manuscript and Dr A. Bell for valuable advice on the fungal group. All are members of the Botany Department, Victoria University of Wellington.

Kingdom Monera

This comprises the bacteria, bacteria-like organisms and blue-green algae. The Monera are the only group to have procaryotic cells. These do not have a membrane-bound nucleus, mitochondria or plastids.

Plate 1

X 2,200

Nitrogen-fixing bacteria *(Rhizobium trifolii)* in root nodule cells of white clover (*Trifolium repens* L.)

Various species of the bacterial genus *Rhizobium* enter the roots of many legumes (members of the family Leguminosae) via the root hairs (Plate 118). They cause the formation of warty swellings, known as nodules, on these roots. The rod-shaped bacteria multiply to fill the cytoplasm of many cells within the root, as shown below. The bacteria are surrounded singly and in groups by parts of the cell membrane of the host cell and are transformed into swollen, sometimes branched, forms called bacteroids. Higher plants do not have the ability to assimilate atmospheric nitrogen, but these bacteria are able to acquire this nitrogen and eventually it is made available to the plant. The process whereby gaseous nitrogen is converted into organic nitrogenous compounds is known as nitrogen fixation. Nodulated leguminous plants are therefore able to grow well in soils which are deficient in nitrogen and are of considerable importance in agriculture.

Rhizobium must be within the cells of the leguminous plant before nitrogen fixation can occur. Such an association between dissimilar organisms to their mutual advantage is known as symbiosis. The biochemistry of symbiotic nitrogen fixation is not fully known.

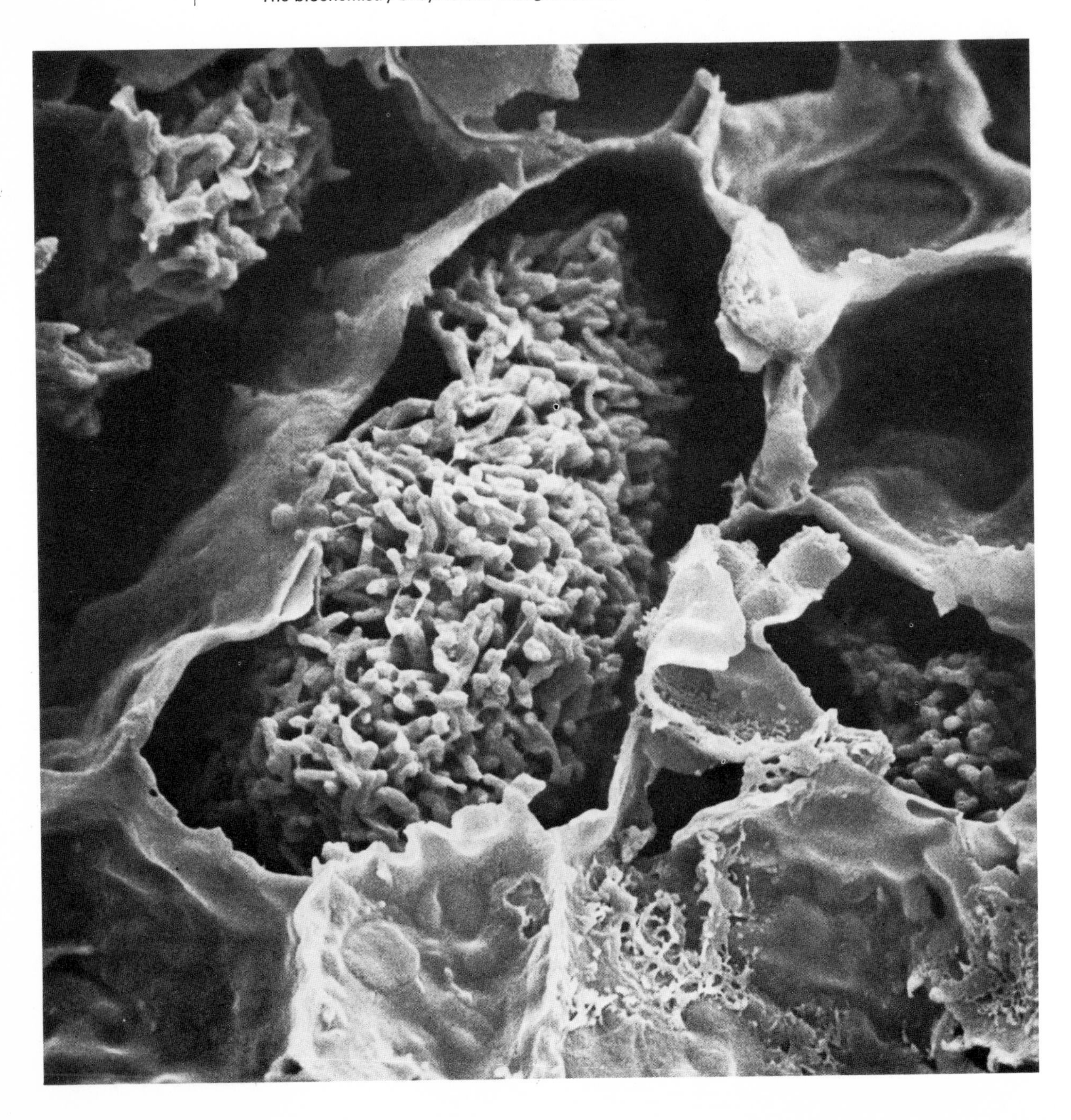

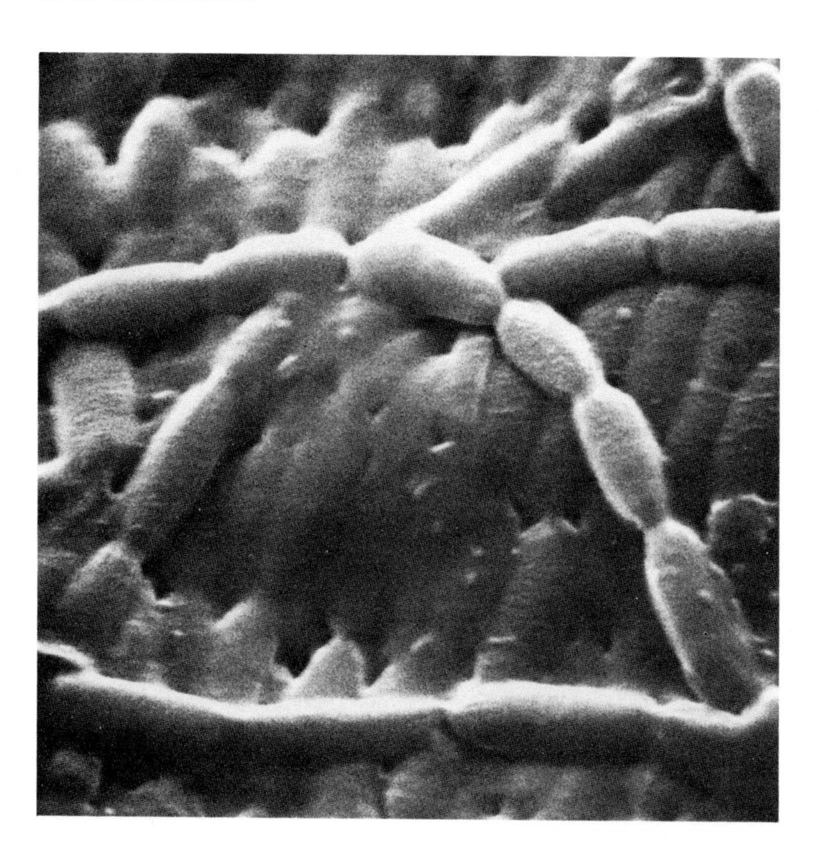

Plate 2 X 7,300

Bacillus bacterium

The *Bacillus* bacteria are rod-shaped. The above species is aerobic (lives in the presence of free oxygen). Although bacteria are unicellular, individual cells may link together to form chains, as illustrated above. *Bacillus* bacteria are among the most common organisms to appear when soil is placed on agar plates containing nutrient media. *Bacillus* and *Clostridium* (Plate 3) have endospores. These are spores which are produced within the cell and are remarkably heat resistant. One species of *Bacillus* causes anthrax disease in farm animals and this is occasionally transmitted to humans. Many bacilli are insect pathogens and toxins from the bacteria are becoming important for use in insecticides. Some *Bacillus* species are used for the production of antibiotics.

Plate 3 X 25,000

The bacterium *Clostridium welchii*

This is a comparatively large bacterium of the bacillus type. Heat-resistant strains of this species are responsible for a type of food poisoning when infected meat is eaten. *Clostridium* is unable to grow in the presence of air and is thus one of the anaerobic bacteria. The bacteria responsible for tetanus and botulism are species of *Clostridium*.

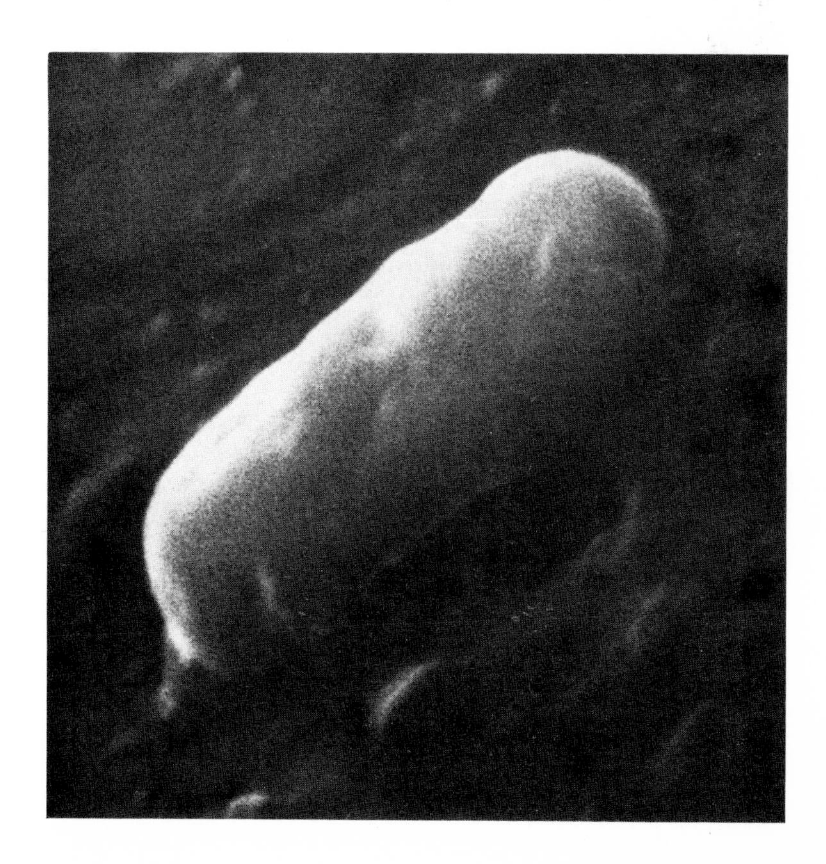

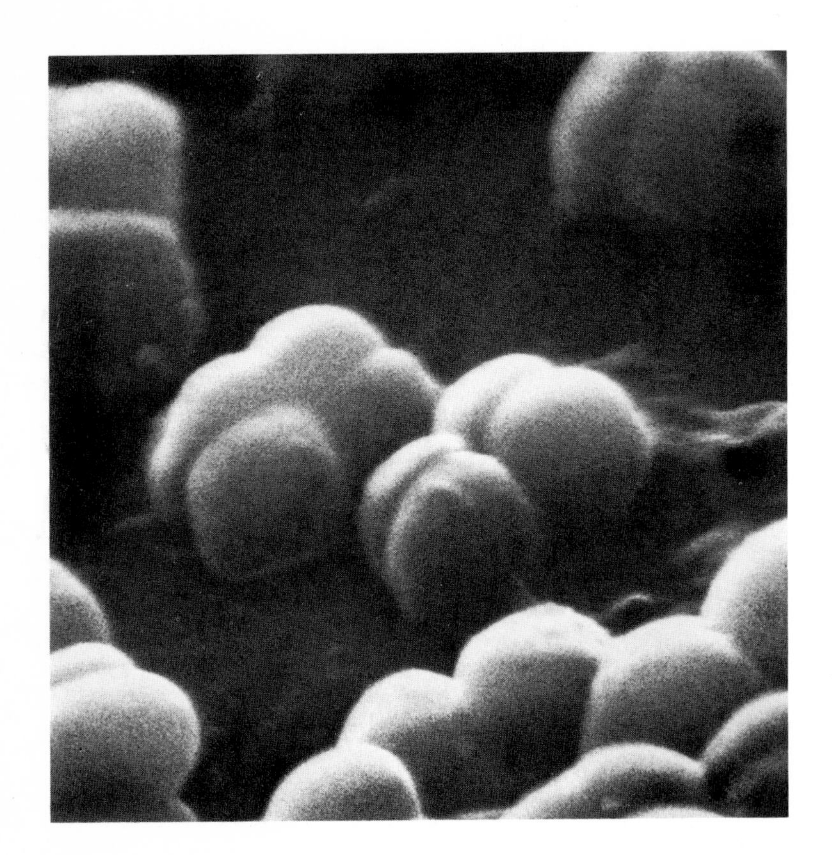

Plate 4

X 17,000

The bacterium *Micrococcus radiodurans*

This bacterium of the coccus (spherical) type is extremely resistant to atomic radiation. The dosage of radiation required to destroy it is approximately 750 times the lethal dosage for humans. The bacteria are grouped in clusters of four cells.

Kingdom Protista

This comprises unicellular organisms or those in which cells are grouped into a simple filament or colony. They have membrane-bound nuclei and mitochondria.

PHYLUM CHRYSOPHYTA

These members of the Protista Kingdom are autotrophic (manufacture their own organic matter by photosynthesis). They possess the photosynthetic pigments chlorophyll *a* and *c* and fucoxanthin.

CLASS BACILLARIOPHYCEAE

The diatoms. They have double siliceous shells, which fit together like the halves of a petri dish.

Plate 5

X 2,300

The diatoms *Melosira*

This is a fresh water species of *Melosira*. Diatoms differ from other algae in having a highly silicified cell wall (containing silica, SiO_2). Each cell is composed of two overlapping halves called valves. The wall of an entire cell is known as a frustule and frustules are united to form chains, which have become broken in the material illustrated. The terminal valve in a chain bears spines at its exposed end and several of these valves can be seen in the photo. When a chain becomes broken, two valves of different (adjacent) frustules frequently remain in pairs. For example, the pair of valves at **A** are from adjacent frustules. Each has a tapered end which would fit into a valve with a wider end, e.g. the one at **B**. The cell walls have regularly arranged pores in them and, in some valves, pores are hidden by surface material.

The siliceous wall is formed in a membrane-bound vesicle, the silica-lemma, within the outer cytoplasm of each new cell. It is then moved to the outside (Stoermer, Pankratz and Bowen, 1965).

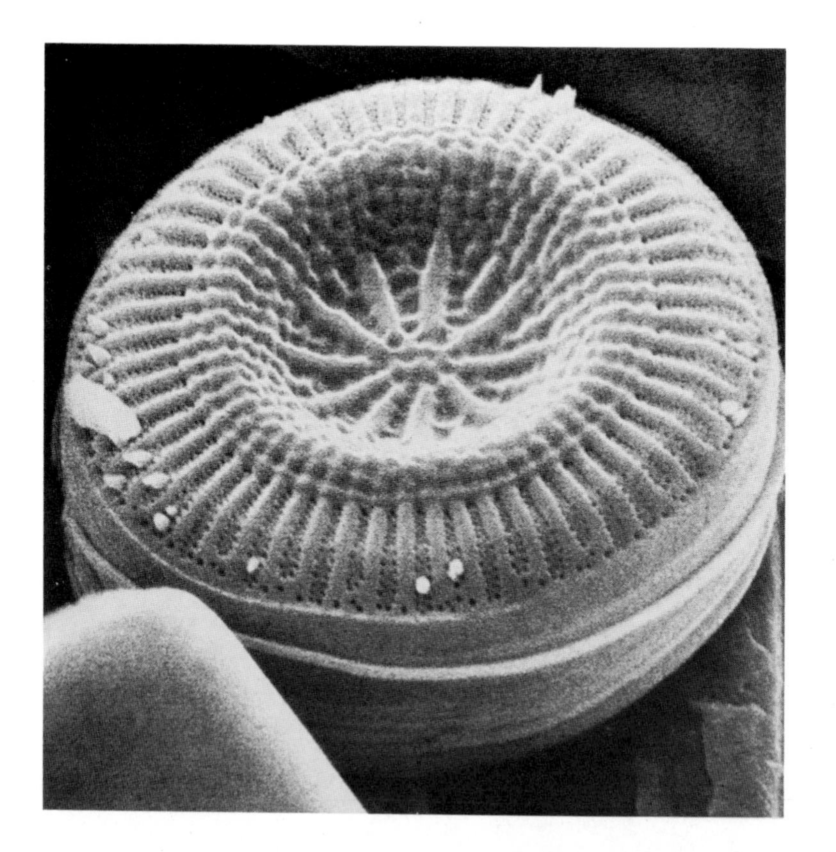

Plate 6 X 5,000

Unidentified diatoms

Plates 6, 7 and 8 illustrate some other diatoms which were not identified.

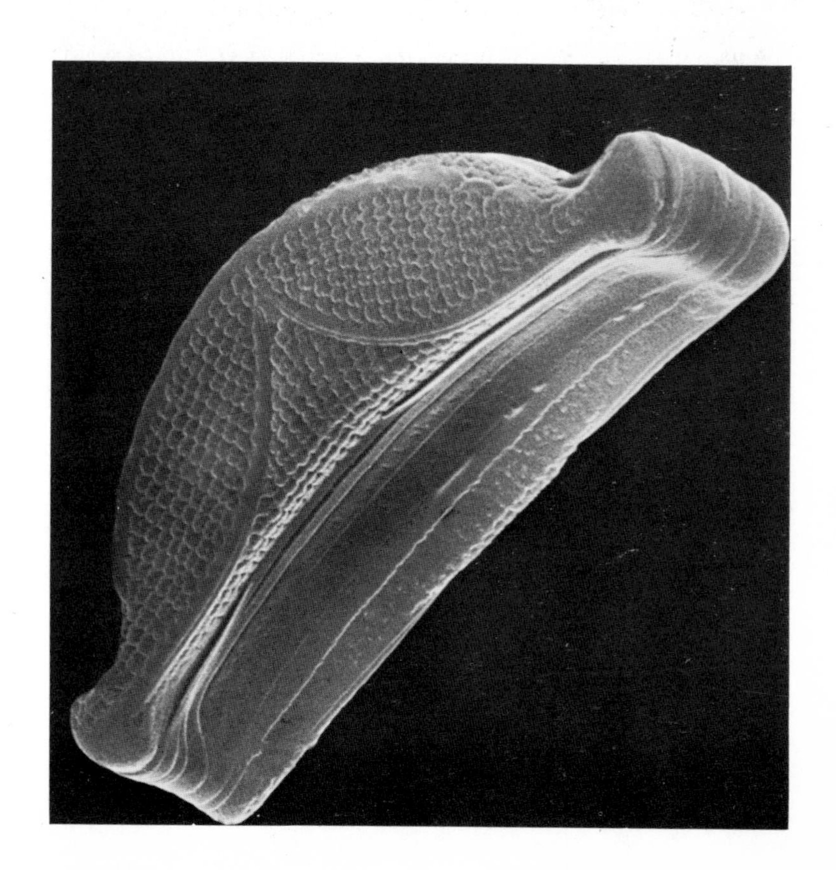

Plate 7 X 3,000

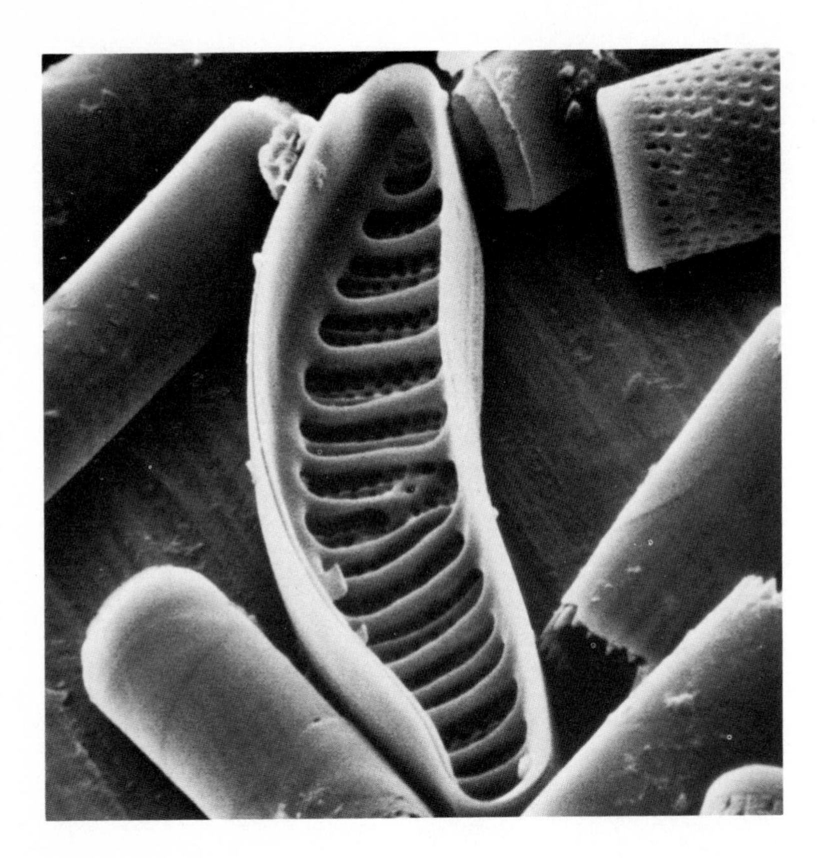

Plate 8

X 3,200

Unidentified diatoms

See Plates 6 and 7.

Kingdom Fungi

The members of this group are heterotrophic (i.e. unable to manufacture their own organic material). They have nuclei which occur within a more or less continuous mycelium, which is, however, septate in some groups and septate at certain stages in the life-cycle of others.

PHYLUM MYCOTA

At present all fungi are placed in this group.

Plate 9

X 480

Fungal hyphae in apple (*Malus pumila* Miller)

This is a section of an apple fruit that has been infected by one of the fungi which cause brown rot of apples. The fungal hyphae form a ramifying network of threads through the apple parenchyma cells. Unfortunately the fungus was not identified when the photo was taken.

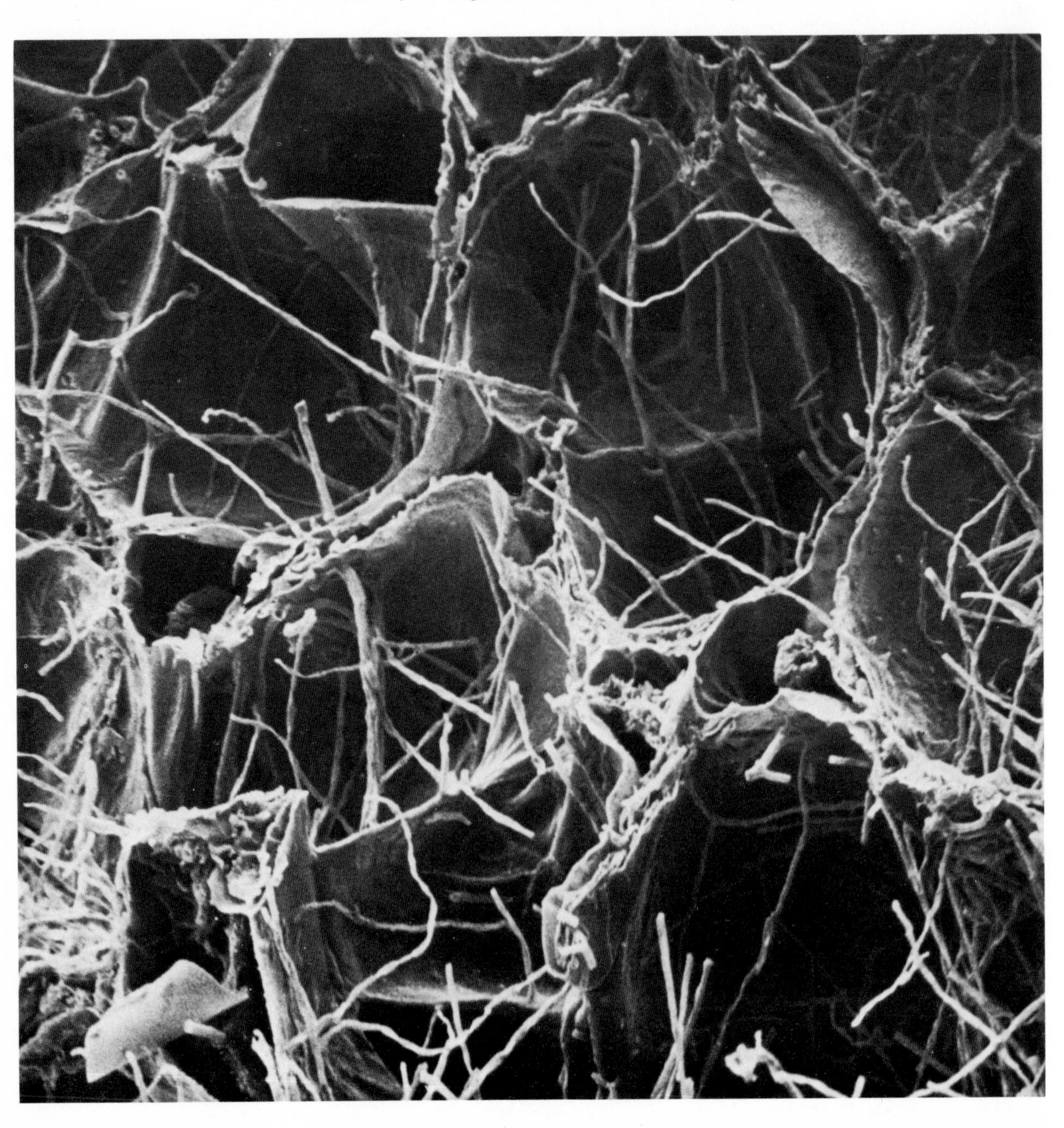

CLASS OOMYCETES

These are characterised by reproducing asexually in most species by means of biflagellate zoospores and by forming oospores in all but the most primitive species.

Plate 10

X 140

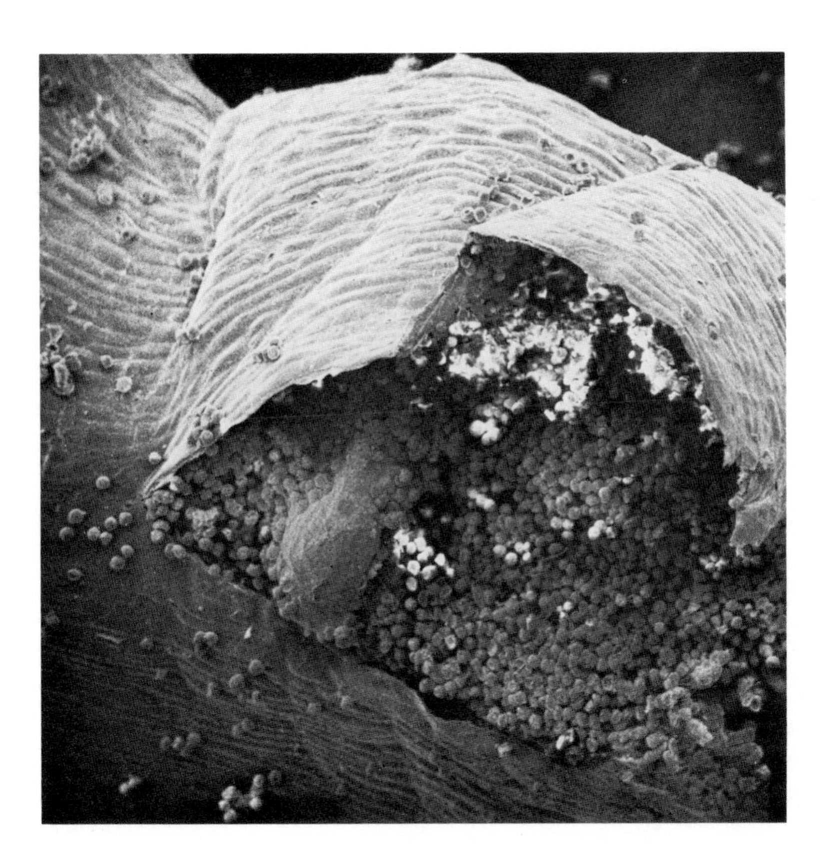

White rust (*Albugo candida* (Hook.) O. Kuntze) **on the stem of shepherd's purse** (*Capsella bursa-pastoris* Medic.)

This cosmopolitan species forms a white rust on the aerial parts of many members of the mustard family (Cruficerae) including shepherd's purse. Non-septate hyphae of *Albugo* penetrate host tissues and in certain regions the hyphae develop in great abundance, just beneath the host's epidermis. Numerous short, erect, club-shaped branches, the sporangiophores, are formed here and at their tips chains of sporangia are cut off. As they mature, the sporangia become detached and accumulate in the space between the epidermis of the host and the sporangiophores. Growth of the fungus finally ruptures the epidermis, as shown in the photo. The sporangia, which are multinucleate, are dispersed by wind and water. If temperatures are suitable, the sporangia germinate to form motile biciliate zoospores which, when they come to rest, develop a germ tube which is able to penetrate host tissue and start a new mycelium.

Sexual reproduction also occurs in *Albugo*. Male and female organs develop on separate hyphae within host tissue (consult Alexopoulos, 1962).

CLASS ZYGOMYCETES

These are saprophytic or parasitic fungi and most have hyphae in which cross walls are formed only during the development of reproductive bodies. Some members have septate hyphae. Zygomycetes are characterised by the production of a sexual resting stage called a zygospore, which typically results from the complete fusion of two gametangia. A gametangium is a structure which contains gametes.

Plate 11

X 550

Sporangium of *Mucor fragilis*

Mucor grows on decaying animal and vegetable material. Asexual reproduction is by spores, which are produced in sporangia. Each sporangium is borne on a stalk, the sporangiophore (**S**), which arises from the fungal mycelium. Note the sculpturing on the wall of the sporangium. Many multinucleate spores form within the sporangium and are liberated by dissolution of the sporangial wall. They germinate to form new hyphae.

A stage in sexual reproduction in the Mucorales is illustrated for *Zygorhynchus* (Plates 12, 13).

Zygospore of *Zygorhynchus* Vuill.

Plate 12

X 1,400

The zygospore (**Z**) is produced as the result of sexual reproduction. The tips of two special branches, called progametangia, come in contact with each other. A wall forms near the tip of each progametangium, separating it into a terminal gametangium, containing cytoplasm and many nuclei, and a suspensor. The walls dissolve where the gametangia touch and the nuclei from different gametangia fuse in pairs to form a number of diploid nuclei. The new cell thereby formed is the zygospore, and it enlarges and develops a thick, sculptured wall. Eventually, after a resting period, the zygospore wall cracks and a sporangiophore emerges and develops a sporangium known as a zygosporangium. Meiosis occurs during zygospore germination and the spores, which are liberated from the zygosporangium, are therefore haploid. They germinate into new hyphae.

Zygorhynchus is homothallic, i.e. the two progametangia arise from the same mycelium. In some of the Mucorales, only progametangia from two separate plus and minus strains of mycelia fuse to form the zygospores. In *Zygorhynchus* the two fusing gametangia and the suspensors are of unequal size. The small suspensor is hidden in the photo beneath the central zygospore (**Z**). The larger suspensor (**S**) can be seen. Parts of older zygospores are also in the photo.

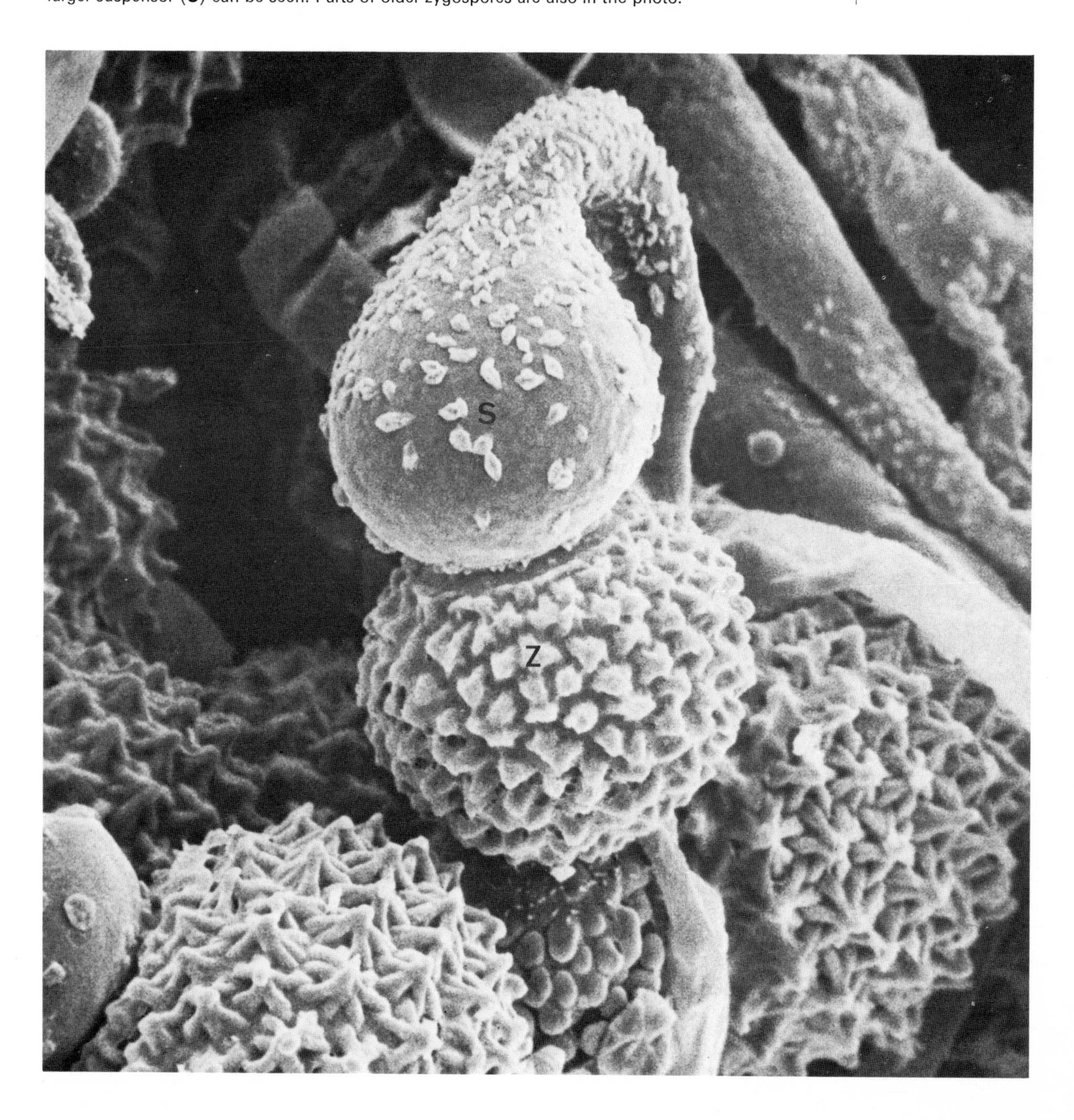

Plate 13

X 5,300

Zygospore wall of *Zygorhynchus*

This is a more magnified view of the wall of a zygospore, which is at a later stage of development than the one in the centre of the previous photo. The fluted pyramidal protrusions on the wall form an attractive pattern.

CLASS ASCOMYCETES

This group is characterised by the presence of an ascus. This is a sac-like structure which contains ascospores. There are usually eight ascospores per ascus. Ascospores are formed as the result of the fusion of two nuclei, followed by meiosis. Hyphae are septate.

Perithecium of *Sordaria fimicola* Ces. and de Not.

Plate 14

X 370

The flask-shaped fruiting body in the photo is a perithecium. It encloses many asci, each of which contains eight ascospores. There is an opening, the ostiole (arrowed), near the tip of the neck of the perithecium. The asci grow and fill the upper part of the perithecium and one of the asci protrudes through the ostiole. It discharges its spores explosively, disintegrates, and other asci repeat this process in succession. Ascospores germinate into new mycelia, which form perithecia under favourable conditions.

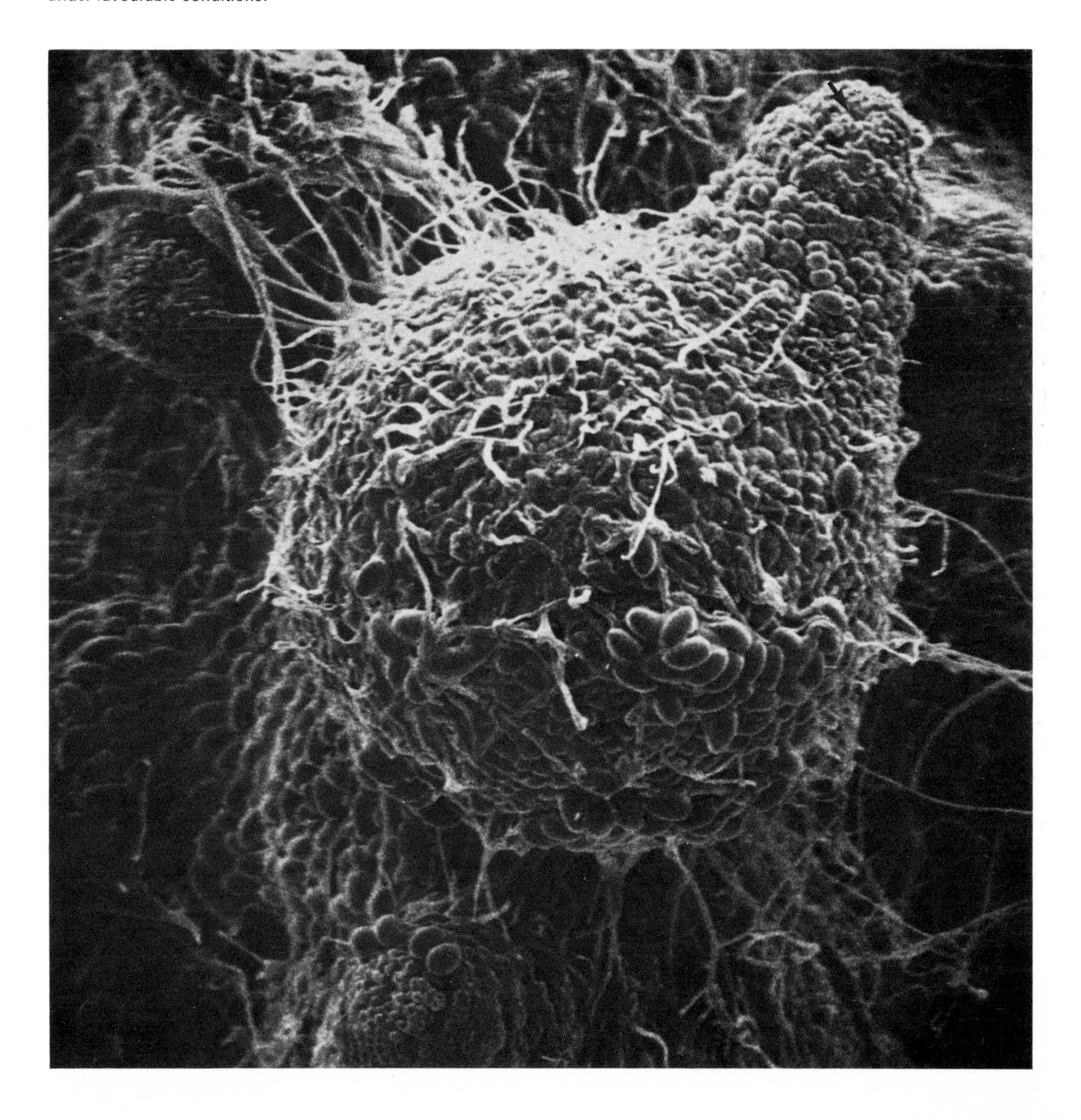

Plate 15

X 900

Ascospores of *Sordaria fimicola*

This is a more magnified view of part of Plate 14, showing some ascospores which have become adherent to the wall of the perithecium. The spores are a dark brown colour and are surrounded by a gelatinous sheath.

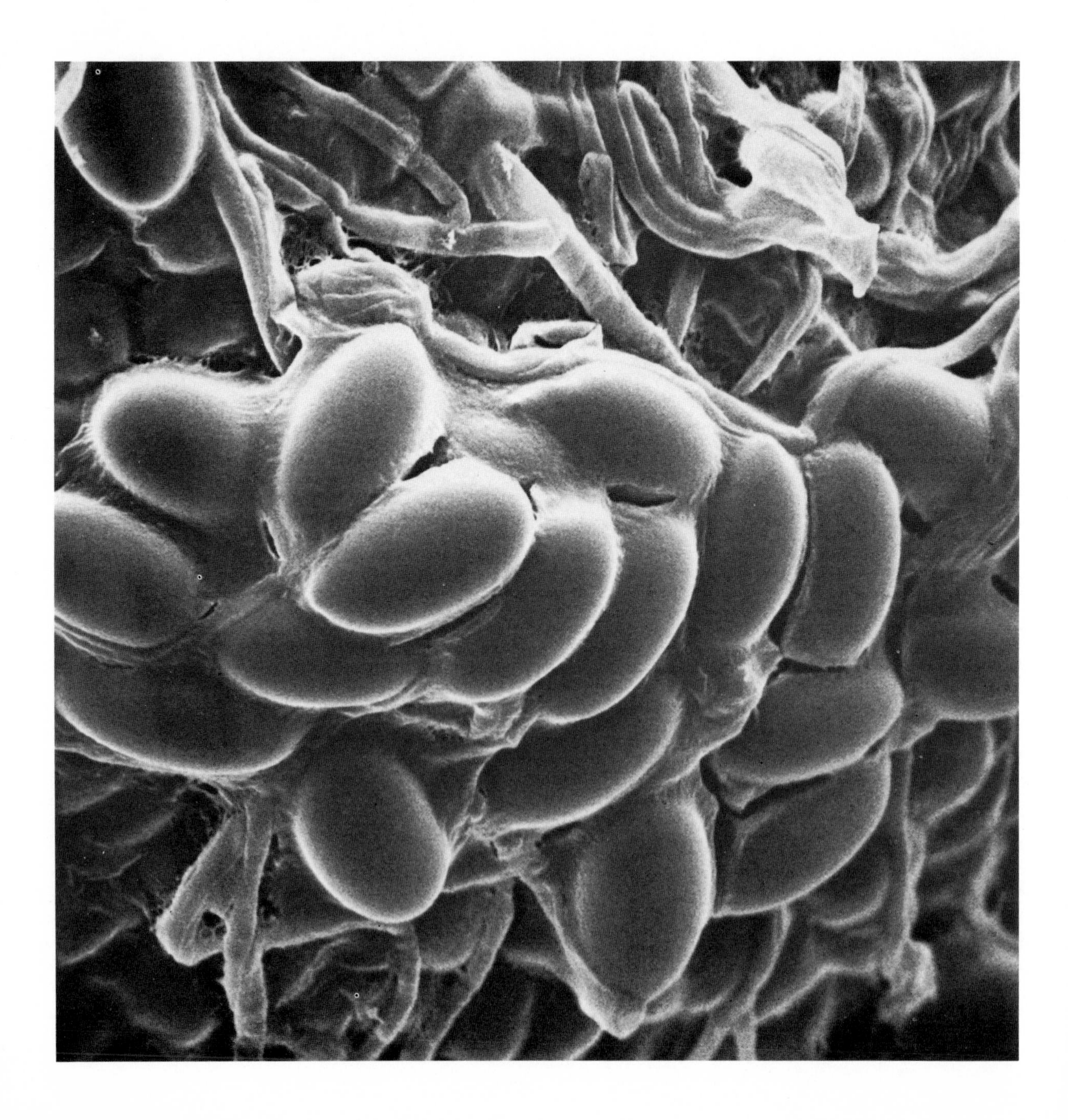

Ascospores of *Ascobolus crenulatus* P. Karst.

Plate 16

X 2,100

Ascobolus grows on dung and soil. The asci occur on the exposed upper surface of a cup-shaped fruiting body, the apothecium. There are eight ascospores in each ascus. The photo shows a cluster of ascospores around what may be the remains of asci. Sculpturing can be seen on the walls of some of the ascospores.

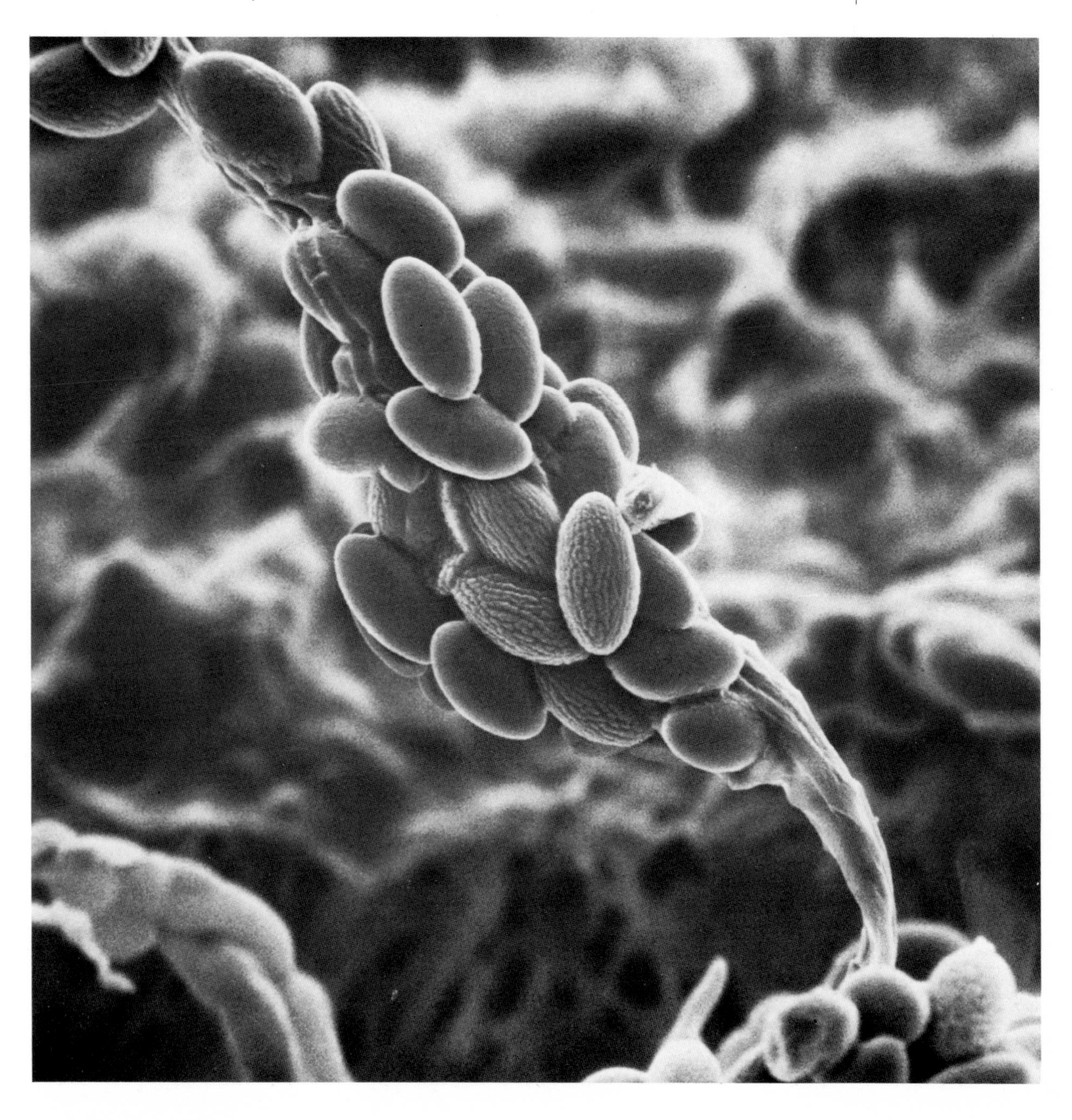

CLASS BASIDIOMYCETES

Members of this group differ from other fungi in that sexual reproduction involves forming specialised structures, the basidia. Meiosis occurs within the basidium and spores are formed on the basidium. Hyphae are septate. During most of their life-cycle, Basidiomycetes are dikaryotic (have two haploid nuclei in each cell).

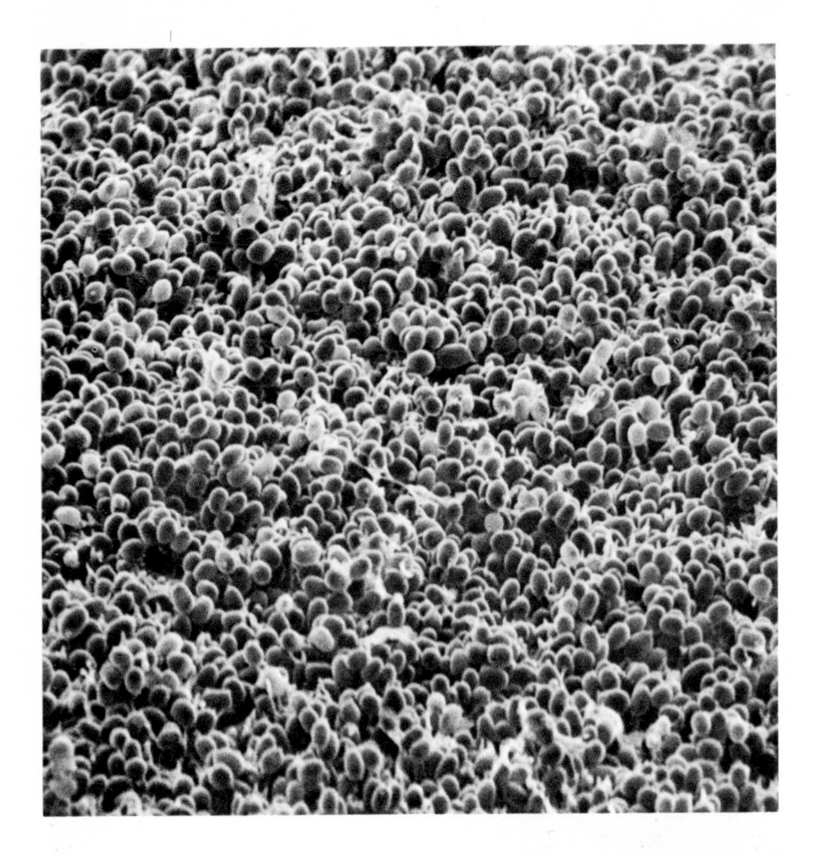

Plate 17 X 600

Gill surface of cultivated mushroom (*Agaricus bisporus* forma *albida* J. Lge.)

The gills are arranged vertically on the underside of the cap of the mushroom. This low magnification photo illustrates the density of spores, which occur on both surfaces of the gills.

Plate 18
X 1,800

Spores and basidia of cultivated mushroom (*Agaricus bisporus* forma *albida*)

This is a more magnified view of part of the gill surface shown in Plate 17. A small protrusion can be seen on many of the spores. This marks the position at which the basidiospore became detached from the basidium. The cultivated mushroom is unusual in having two instead of four spores produced from each basidium, which explains its specific name *bisporus*. Two slender protrusions, the sterigmata (arrowed) visible on basidia which have lost their spores, are the processes where the spores were attached.

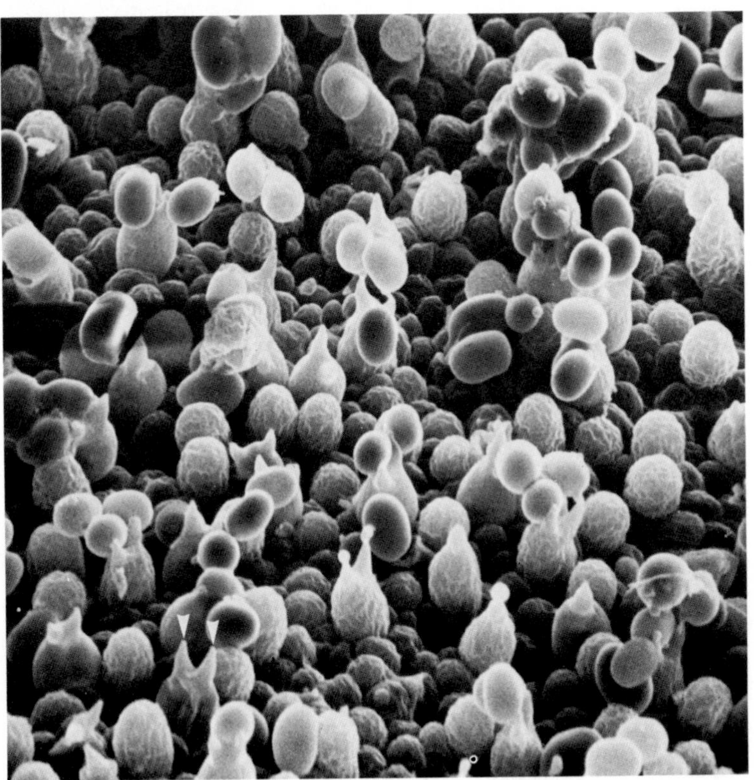

Rust infection (*Phragmidium mucronatum* (Pers.) Schlecht.) on rose leaf

Plate 19

X 580

The rusts (order Uredinales) are Basidiomycetes which are parasitic on plants, and they cause considerable damage to many crop plants. Rusts have a complicated life-cycle consisting of four or five different reproductive stages.

Plates 19 and 20 illustrate the uredospore stage of the life-cycle. A fungal mycelium of binucleate cells grows beneath the epidermis of the leaf and one-celled binucleate uredospores are formed. Each uredospore is on a long stalk and the spores are borne in structures called uredia. The spores rupture the host's epidermis (Plate 19) to form an exposed cluster of uredospores known as the uredosorus. Sterile hairs (paraphyses (**P**)) are intermingled with the uredospores. Stomata can be seen on the rose leaf. Plate 20 shows a later stage in the development of a uredosorus. Sculpturing is visible on the spore walls. The uredospores germinate into binucleate mycelia, which in a few days can form new uredospores. This stage is thus very effective in propagating the fungus.

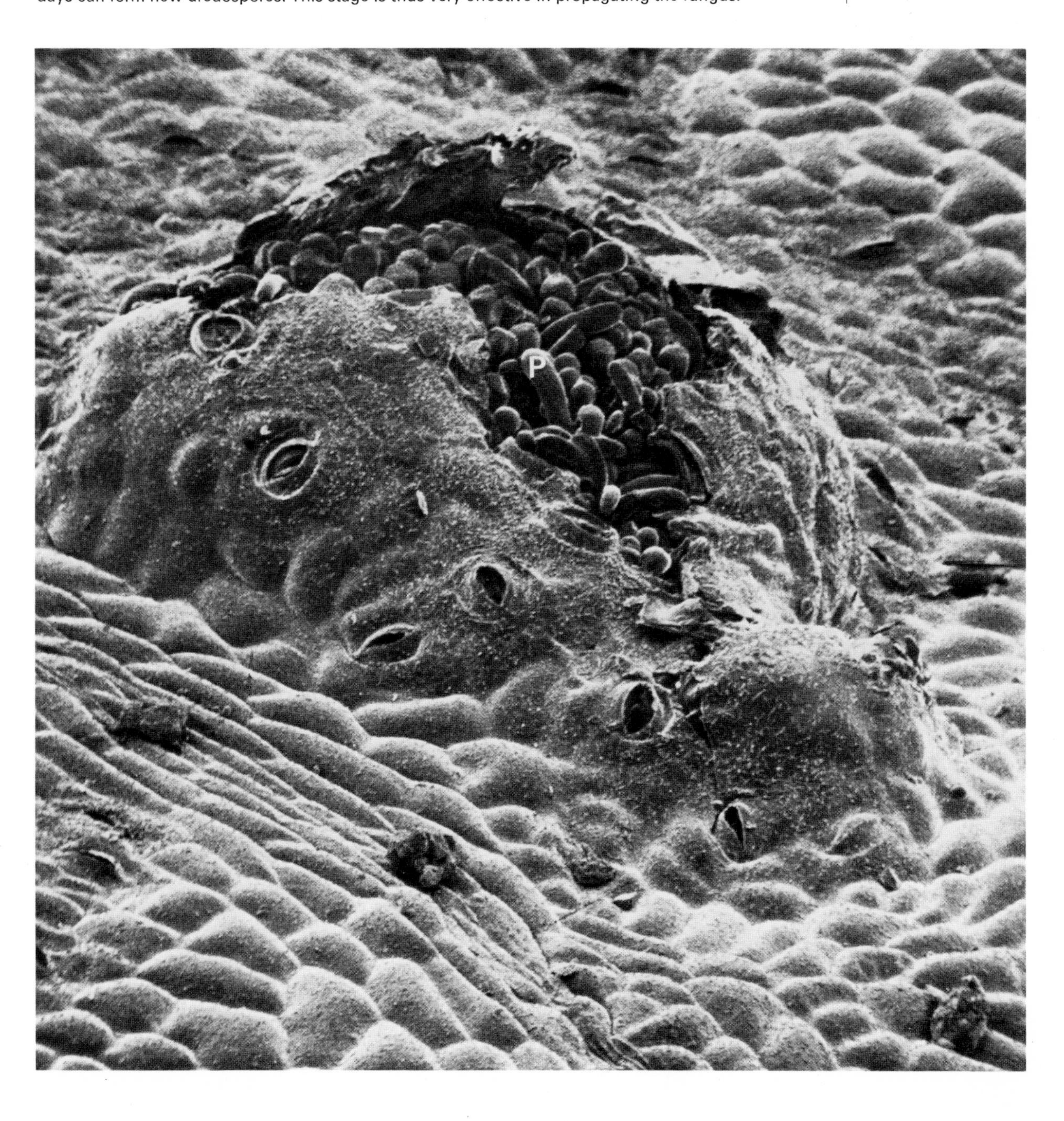

Plate 20
X 470

Later stage in development of rust infection on rose leaf

See Plate 19.

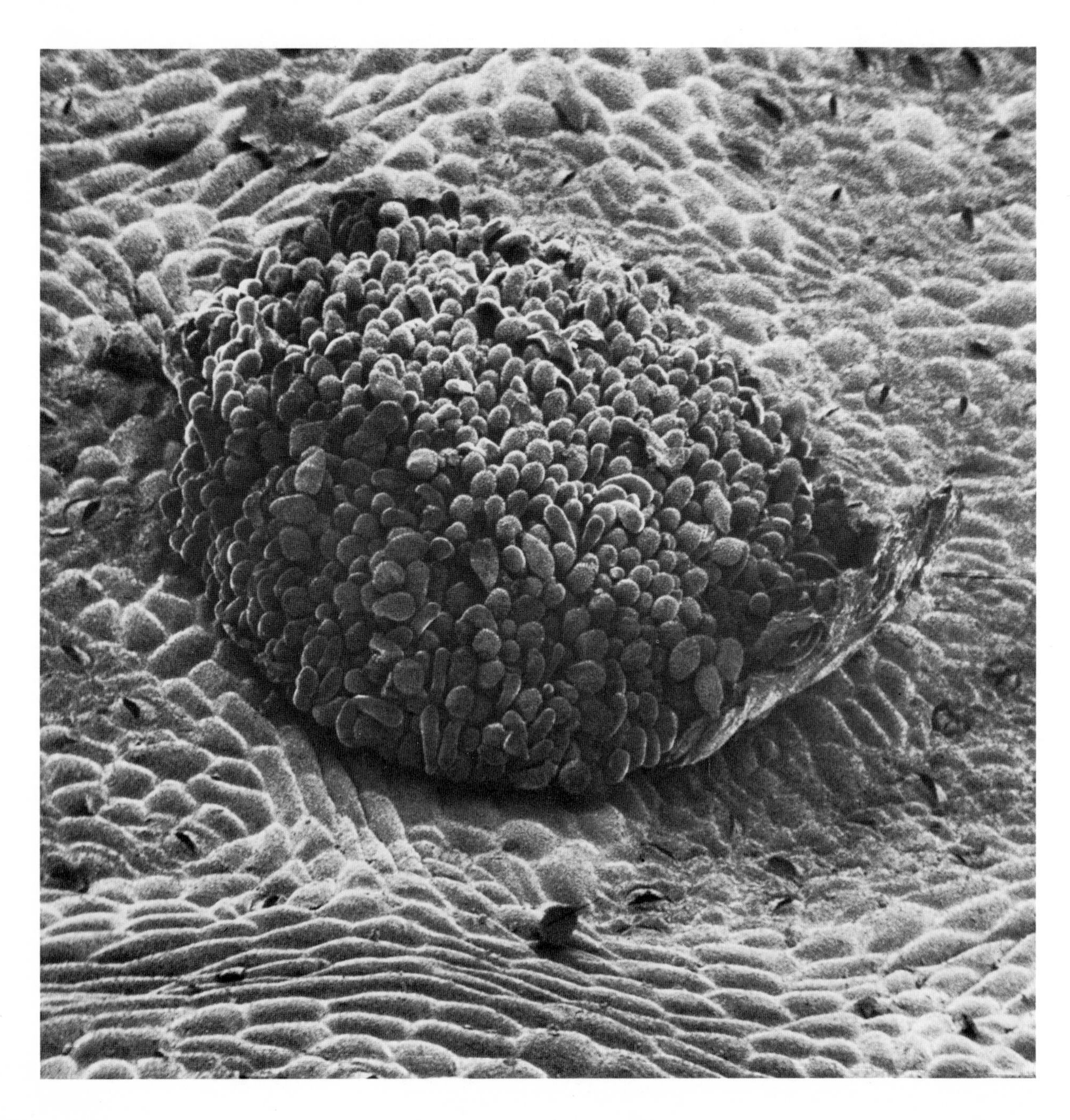

Uredospores of *Phragmidium tuberculatum* J. Muller

This species is also a parasite of rose (see previous plates). Differences in the sculpturing on the uredospore walls are useful for identifying various species of *Phragmidium*.

Plate 21

X 12,000

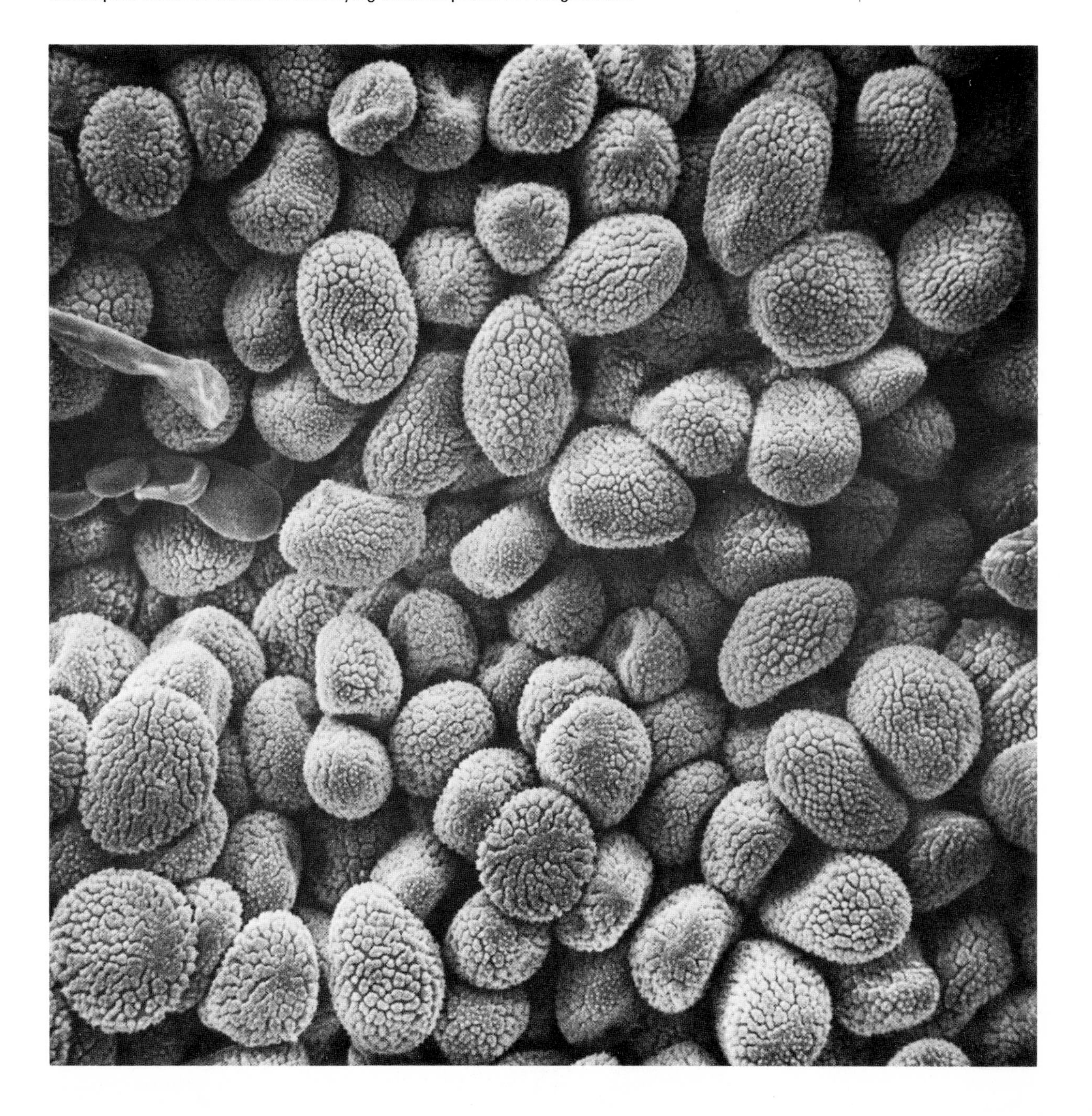

CLASS FUNGI IMPERFECTI

These fungi mainly have the characteristics of Ascomycetes, but sexual stages, leading to the formation of asci, are lacking.

Plate 22

X 3,800

Conidiophores and conidia of *Penicillium* Link. ex Fr.

Penicillium is a cosmopolitan fungus, forming green and blue moulds. The conidia (non-sexual spores) are widely distributed in the air and soil. Conidia (**C**) occur as chains of spores at the ends of specialised hyphae, the conidiophores (**CP**). The conidia are cut off in succession from the apex of a specialised terminal cell on the conidiophore, known as a phialide. When the conidia are released and germinate on a suitable substrate, new hyphae are formed. Further conidiophores are formed on these hyphae. This is a very efficient method of reproduction and enormous quantities of conidia are produced. Conidiophores and conidia have become shrunken because of the partial vacuum which is required for SEM examination.

Stages leading to the formation of asci are lacking in most species. These species are therefore placed in the Fungi Imperfecti. A few species of *Penicillium* do have a sexual stage which leads to the development of asci and they are therefore classified within the Ascomycetes.

Some species of *Penicillium* are used for the manufacture of certain antibiotics including penicillin, and for flavouring certain types of cheese. For further details, consult Alexopoulos (1962).

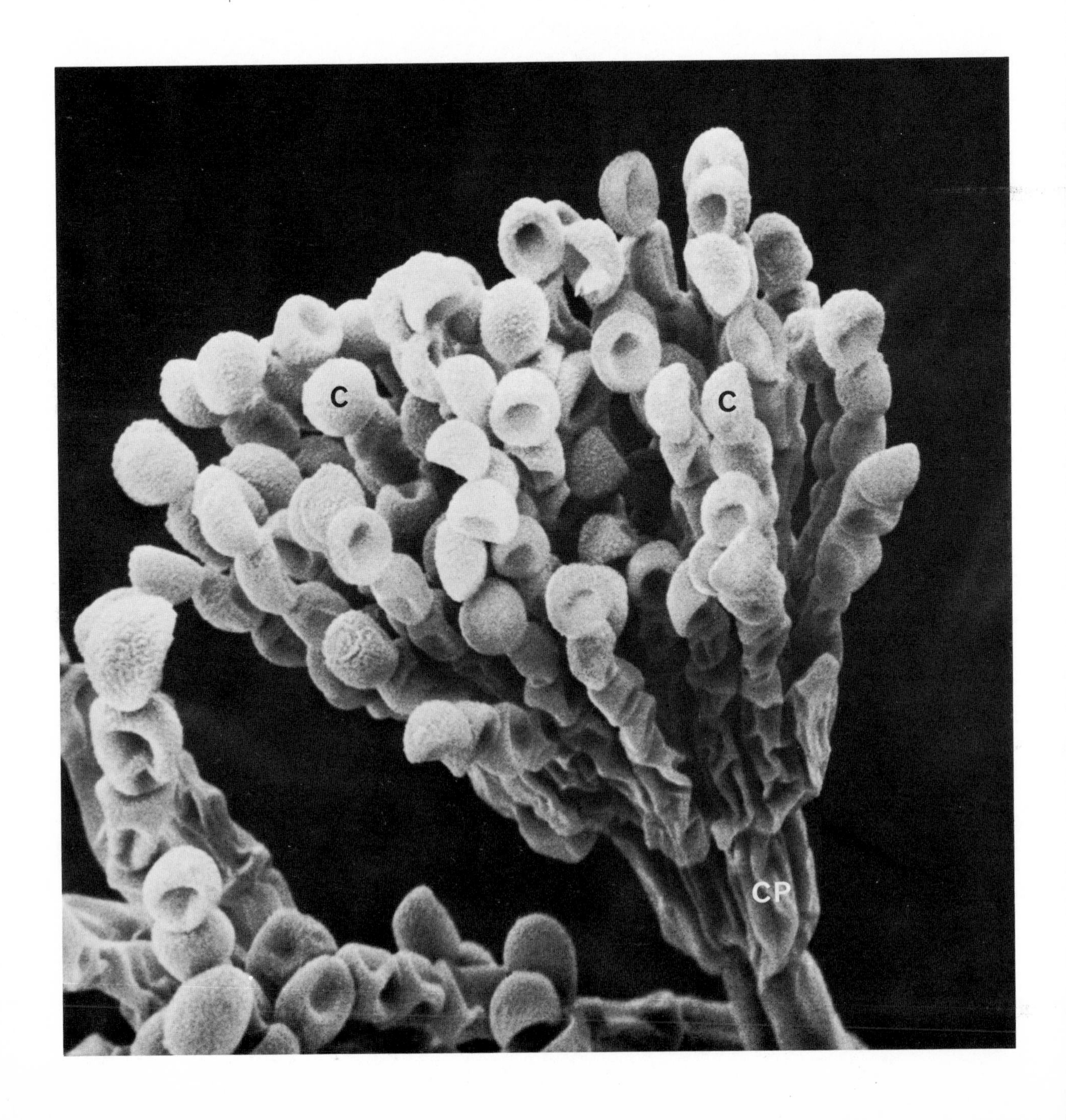

CLASS LICHENS

Dual organisms formed by a symbiotic association of a fungus with an alga. The fungal partner is usually an Ascomycete but sometimes a Basidiomycete. The algal partner is a member of the blue-green algae (phylum Cyanophyta of the Kingdom Monera) or green algae (phylum Chlorophyta).

Sticta latifrons— transverse fracture of a lichen

Plate 23

X 870

A lichen is a symbiotic association of a fungus and an alga in which the two types of plants live together for mutual benefit. The alga photosynthesises enough organic material for itself and the fungus, while the latter provides protection for the alga. The resulting lichen has a form all its own and possesses unique structures such as soredia (reproductive particles composed of a few algal cells and fungal hyphae) and cyphellae (Plate 24). Lichens absorb some minerals from their substrate but many elements are absorbed from the air and in rainfall.

Sticta is a foliose (leafy) lichen with a flat sheet-like thallus growing close to the substrate and attached to it via a short basal stalk in this species. In cross-section this type of foliose lichen has four layers: (i) an upper cortex consisting of heavily gelatinised fungal hyphae forming a protective surface; (ii) an algal layer which also contains loosely interwoven thin-walled fungal hyphae; (iii) the medulla comprising a thick layer of loosely packed, weakly gelatinised hyphae; and (iv) the lower cortex which resembles the upper cortex.

Lichens can grow in environments in which no other form of plant life can survive. An important factor in their survival during very dry conditions is their ability to dry out rapidly. The lichen survives these extremes of heat or cold in a state of "suspended animation" (Raven and Curtis, 1970).

A—algal cell

Plate 24
X 360

Sticta latifrons—cyphella on underside of thallus

Only *Sticta* has the lower surface of the thallus dotted with cyphellae—volcano-shaped empty circular pits, with raised margins. Their function is uncertain although it has been suggested that they facilitate the entry of air into the medulla.

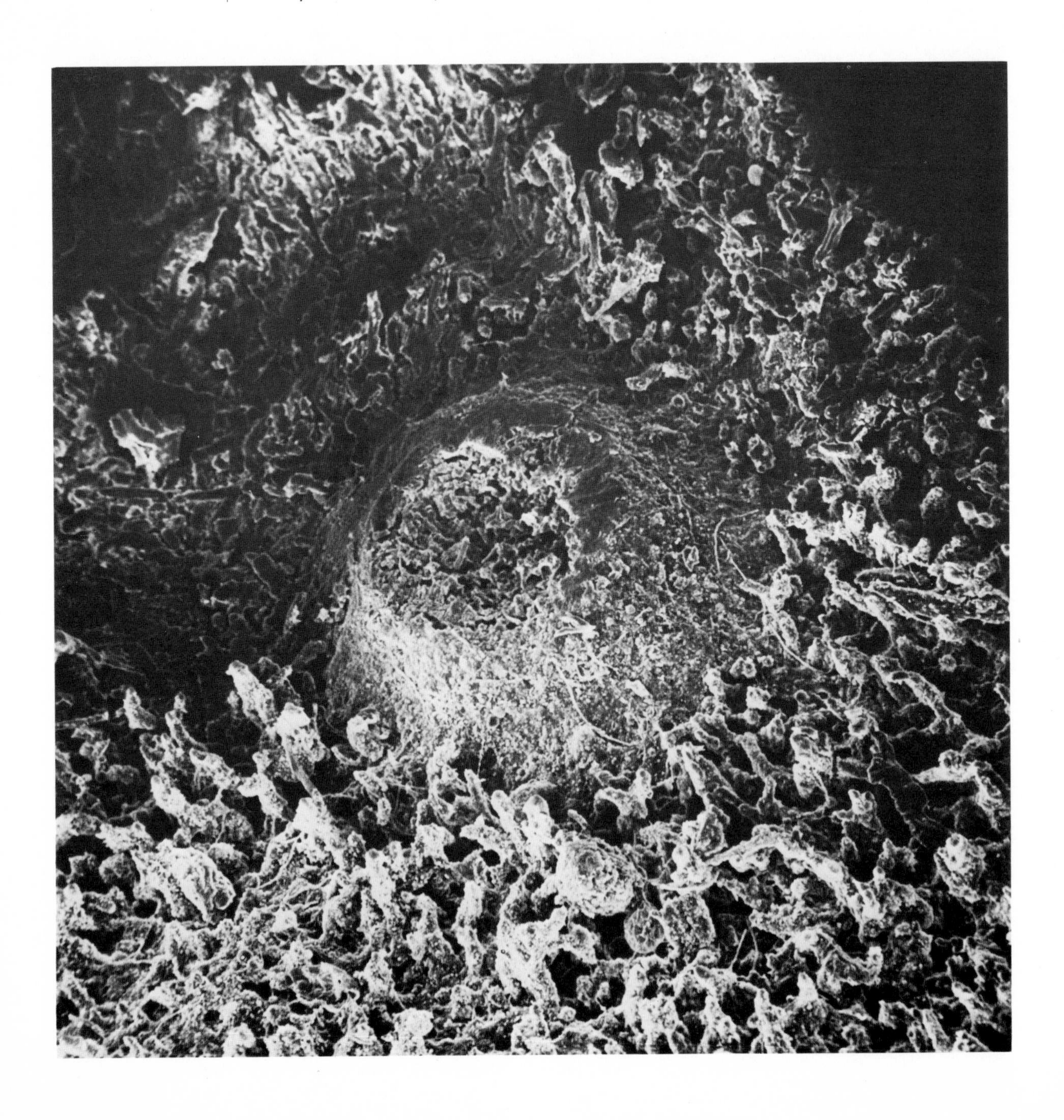

Cladia retipora— **the coral or lace lichen**

Plate 25
X 40

This Australasian lichen occurs in lowland heaths and sub-alpine bogs (Martin and Child, 1972). It is a whitish colour, and the perforated "stems" shown in part in the photo give it an attractive appearance. It belongs to the fruticose group in which the lichen has an erect (or in some cases pendulous) form and is attached at its base to the substrate.

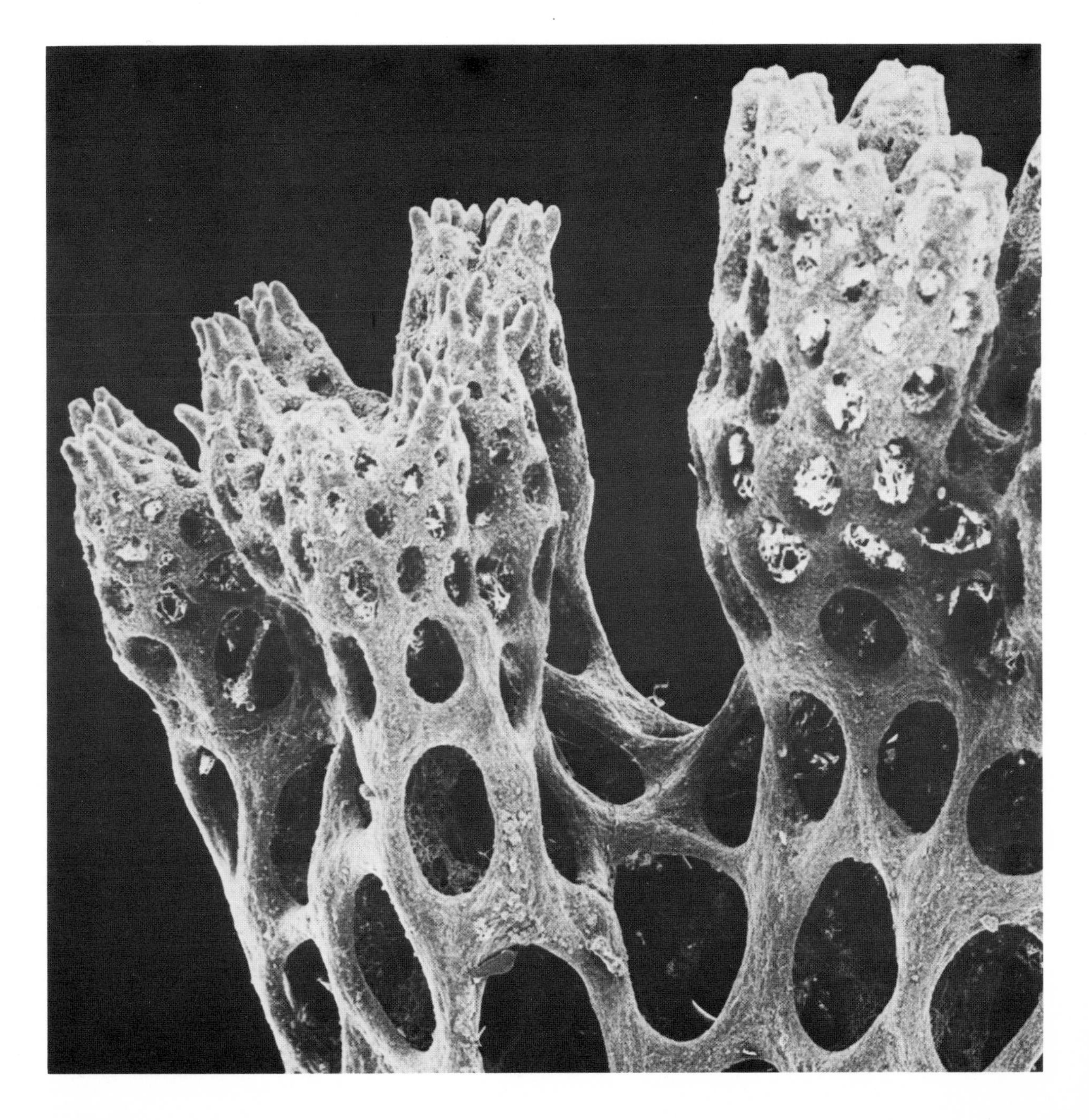

Kingdom Plantae

Generally multicellular organisms with photosynthetic nutrition (a few members have secondarily become heterotrophic).

PHYLUM CHLOROPHYTA

The green algae. These are unicellular or multicellular plants with chlorophyll *a* and *b* and carotenoid pigments. Carbohydrates are stored as starch.

The green alga *Oedogonium*

This fresh water plant consists of cylindrical uninucleate cells which are united end to end to form a filament. The genus is readily recognisable by the transverse ridges situated at one end of certain cells in the filament. Each ridge is formed during cell division (consult Smith, 1955, for details). Other algae are also present in the photo.

Plate 26

X 1,750

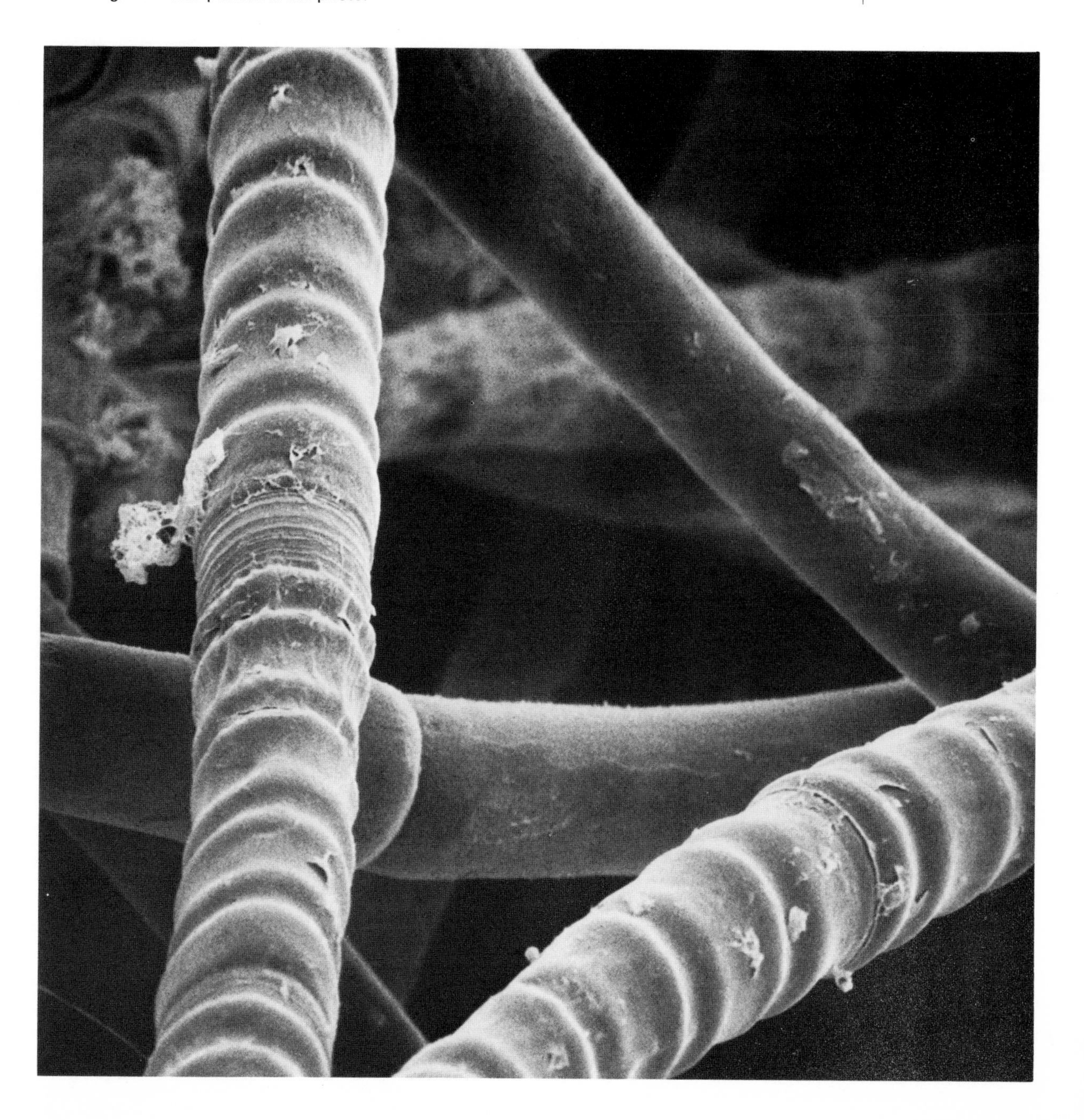

PHYLUM PHAEOPHYTA

The brown algae. These are multicellular seaweeds in which there is considerable cellular specialisation. They are characterised by having the brown pigment fucoxanthin in addition to chlorophyll *a* and *c*. Carbohydrates are stored as laminarin.

Plate 27

X 80

A female conceptacle of the brown seaweed *Hormosira banksii* (Turner) Decne., **in sectional view.**

In the order Fucales, of which *Hormosira* is a member, the sex organs are borne within special flask-shaped cavities in the thallus, known as conceptacles. In *Hormosira* there are separate male and female conceptacles which are on different plants. In the female conceptacle, structures known as oogonia form on the wall of the conceptacle. Four ova are produced within each oogonium by meiosis. Between the oogonia are filaments which cannot be clearly seen in the photo. The oogonia are released through an opening in the top of the conceptacle known as the ostiole (Plate 28) which is not visible in this photo. The ova (female gametes) burst from the oogonia and each ovum is fertilised by a motile sperm, released from a male plant, to form a zygote. This grows into a new plant.

O—ovum

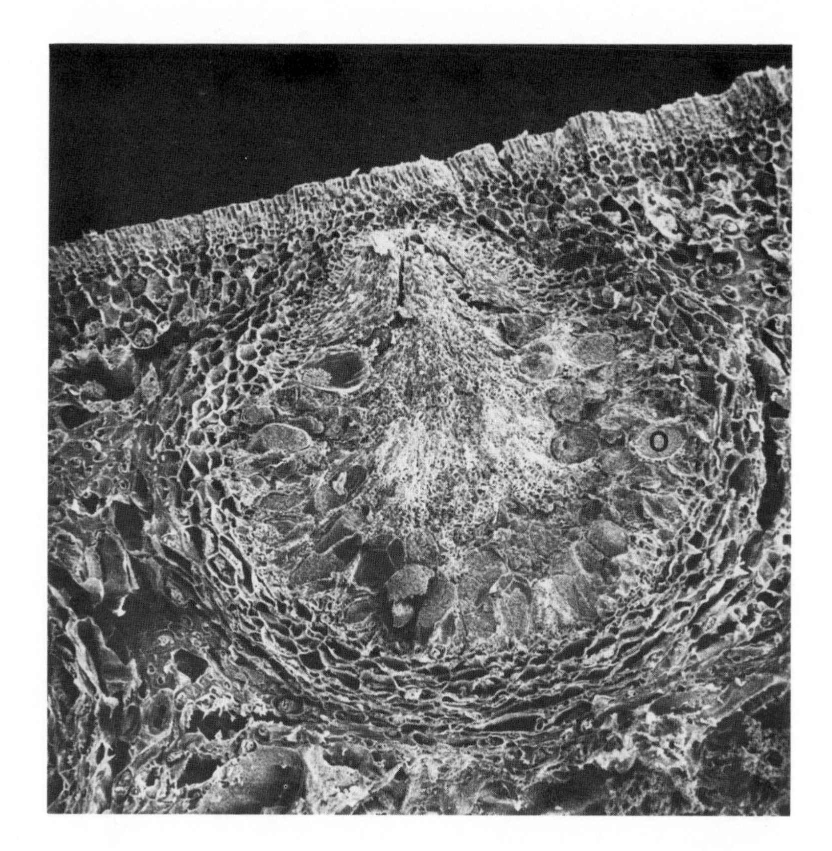

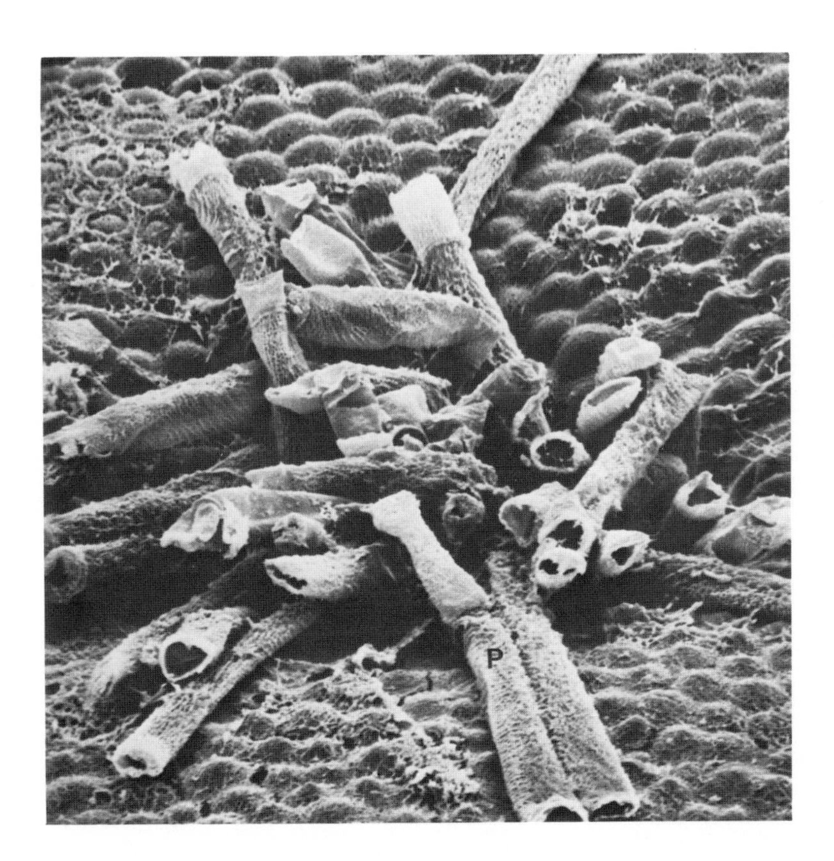

Plate 28

X 700

Ostiole of the brown seaweed
Hormosira

This is a surface view of part of a female plant of *Hormosira* showing filamentous hairs, known as paraphyses (**P**), extending out of the ostiole of a conceptacle. The oogonia (Plate 27) are liberated through the ostiole.

Plate 29

X 5,000

Sieve element and sieve plate of the brown seaweed
Macrocystis pyrifera (L.) C. Ag.

Seaweeds are included among the so-called non-vascular plants, i.e. plants without conducting tissues. Nevertheless a group of brown algae, the Laminariales, possess sieve elements which resemble those in the phloem of highly evolved dicotyledons (Esau, 1969) and there is evidence that food materials are transported in the sieve filaments of these seaweeds (Nicholson and Briggs, 1972).

The sieve elements around the central region of the stipe of *Macrocystis* form a filamentous cellular meshwork and are embedded in mucilage. In Plate 29 part of the wall of a sieve filament, running diagonally from upper right to lower left, has been broken open to reveal part of a sieve plate. The sieve plate represents the end wall between two cells of the filament.

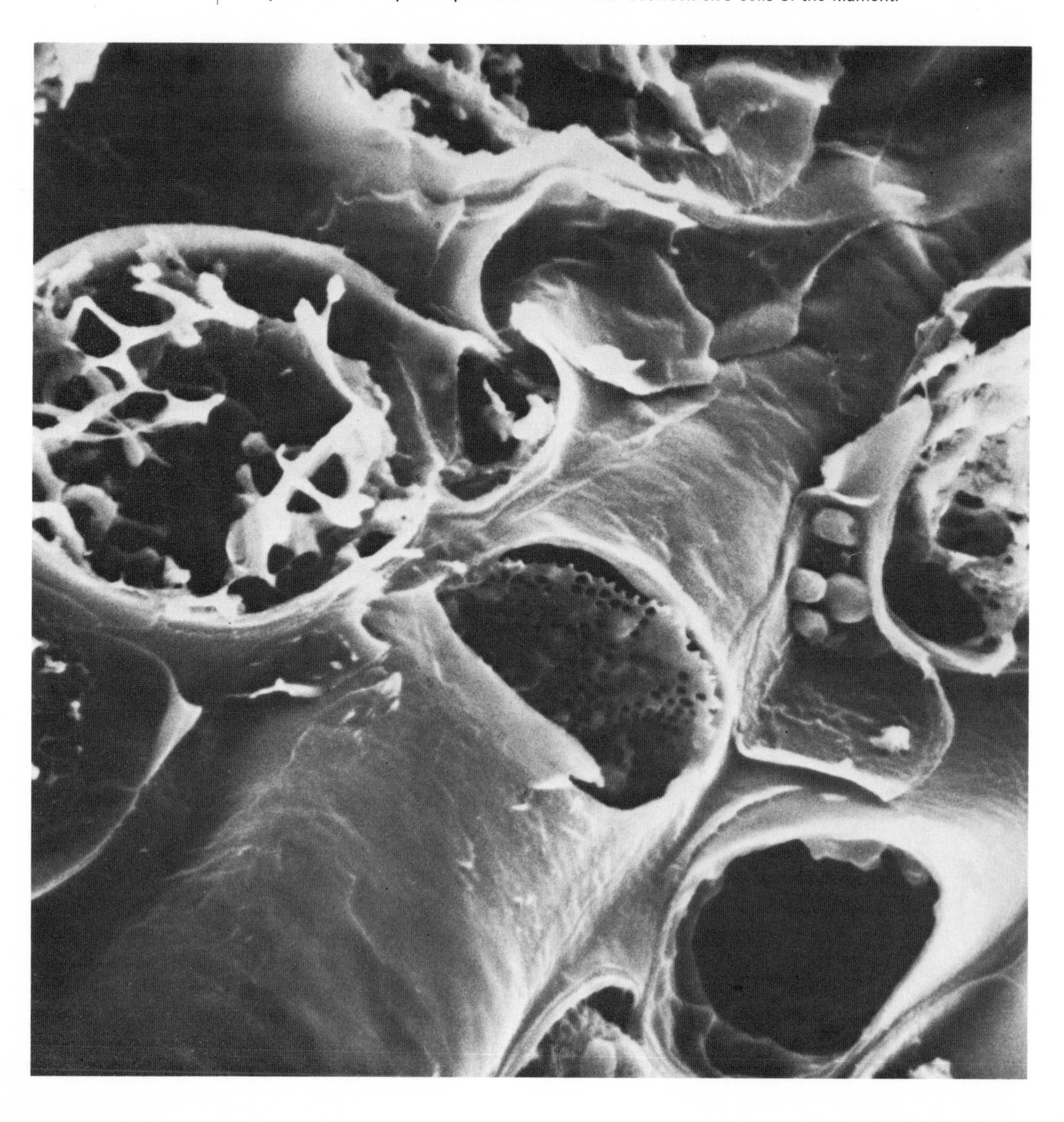

Plate 30 is a more magnified view of another sieve plate. In contrast to pumpkin (Plate 108), pretreatment with sodium hydroxide was not needed to reveal sieve pores. Food materials are transported from one sieve cell to another via these pores, which are lined with callose (not visible).

Plate 30

X 7,000

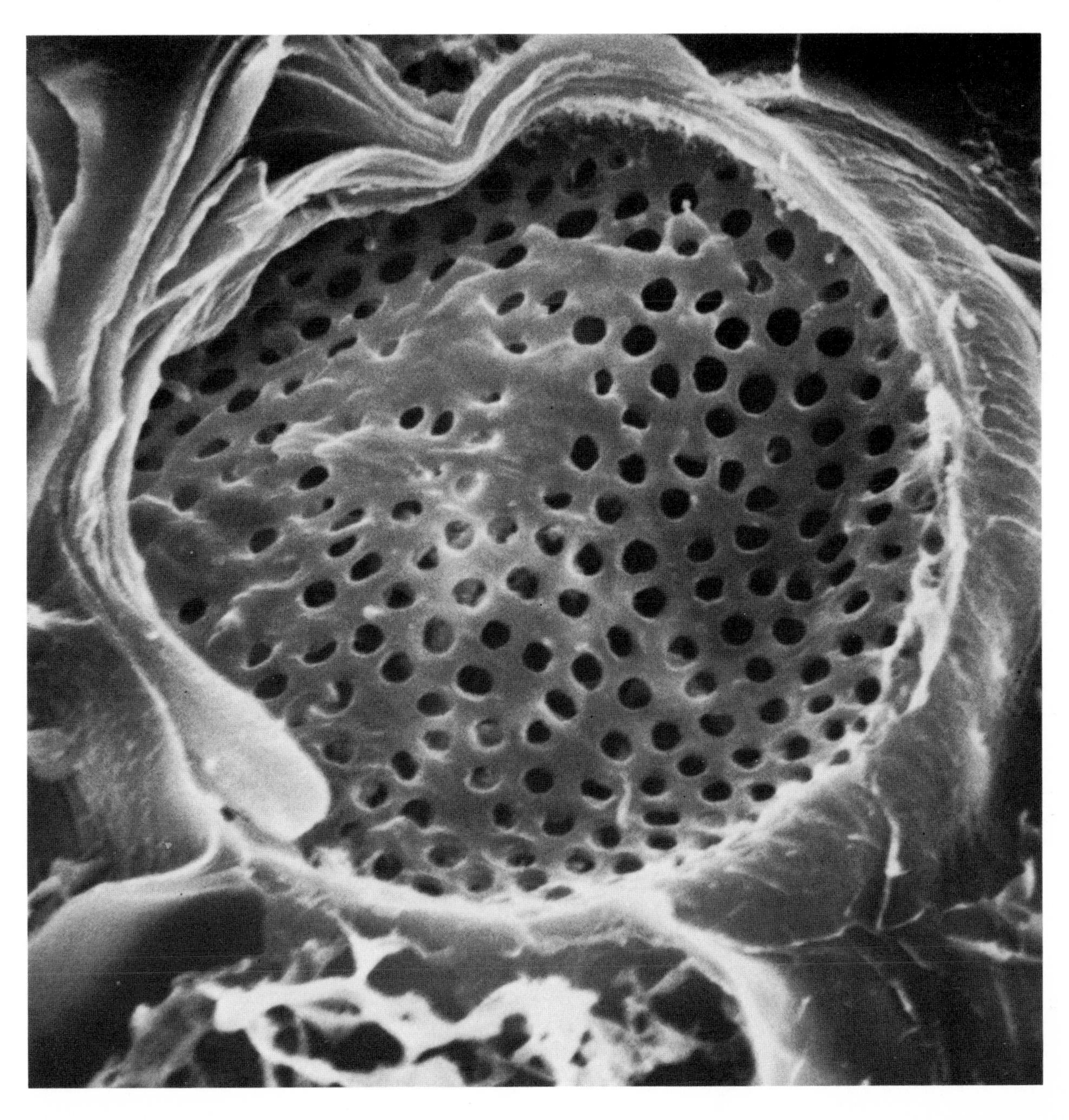

PHYLUM BRYOPHYTA

Liverworts, hornworts and mosses. These are multicellular plants with photosynthetic pigments and food reserves similar to those of the green algae. They have sex organs (antheridia and archegonia) with a sterile jacket layer. Conducting tissue is frequently absent. The conspicuous generation is the gametophyte.

CLASS HEPATICAE

Liverworts. The gametophytes are thallose or leafy with single-celled rhizoids.

Plate 31

X 1,500

Air pore on *Marchantia* thallus

In this southern hemisphere species, *Marchantia berteroana* L. et L., the barrel-shaped air pore consists of usually six tiers of cells, each of three to six cells (Campbell, 1965). In each layer the three to six cells form a ring with an aperture in the centre. The rings lie one above the other. The pore wall lies partly above, partly below, the level of the epidermis. The pores allow gases to pass through the epidermis to the photosynthetic filaments below. Opening of the pore is controlled by the innermost tier of cells; the upper aperture remains the same size. Each cell of the lower tier has an inward projecting part (papilla) which constricts the aperture (not visible in photo). Under dry conditions when the cells lose water and the papillae shrink and flatten, the pore becomes smaller (Campbell, 1965) thereby reducing transpiration. Although the pores act like stomata, a quite different mechanism is involved in their opening and closure.

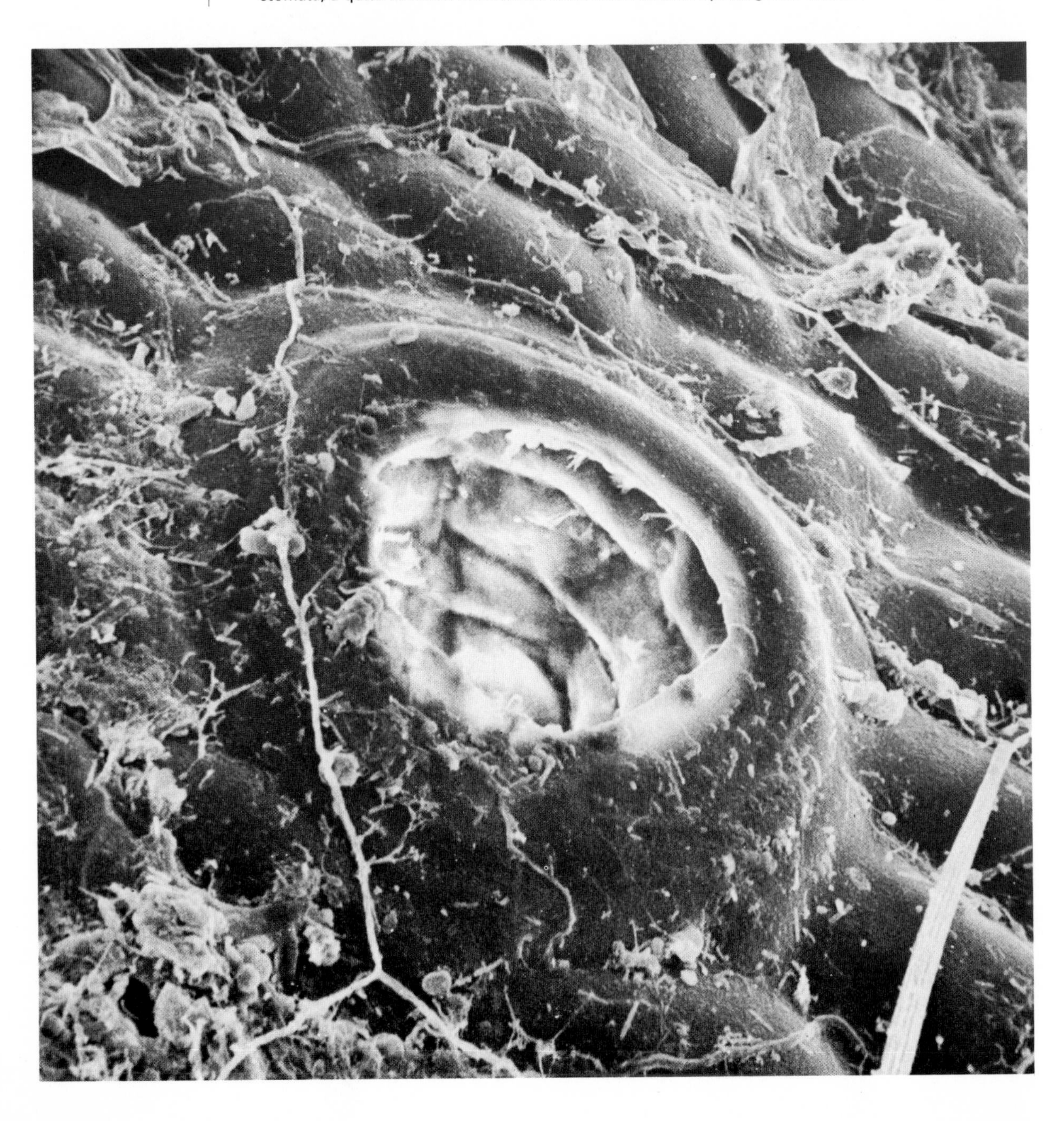

Transverse fracture of upper part of *Marchantia berteroana* **thallus**

At the top are the cells of the upper epidermis (outer walls not visible). They contain few chloroplasts. Beneath this single layer are the photosynthetic filaments, three to four cells high, situated in the air chamber. The filaments contain abundant chloroplasts, which are visible as darker circles within some of the unbroken cells. Free diffusion of gases can occur because the filaments are not packed tightly together. The lower cells in the photo form the flat base of the air chamber. The photo does not illustrate the large-celled ventral tissue, which extends from beneath the base of the chamber to the lower epidermis. It is up to 30 cells in thickness and contains few chloroplasts.

Plate 32

X 2,100

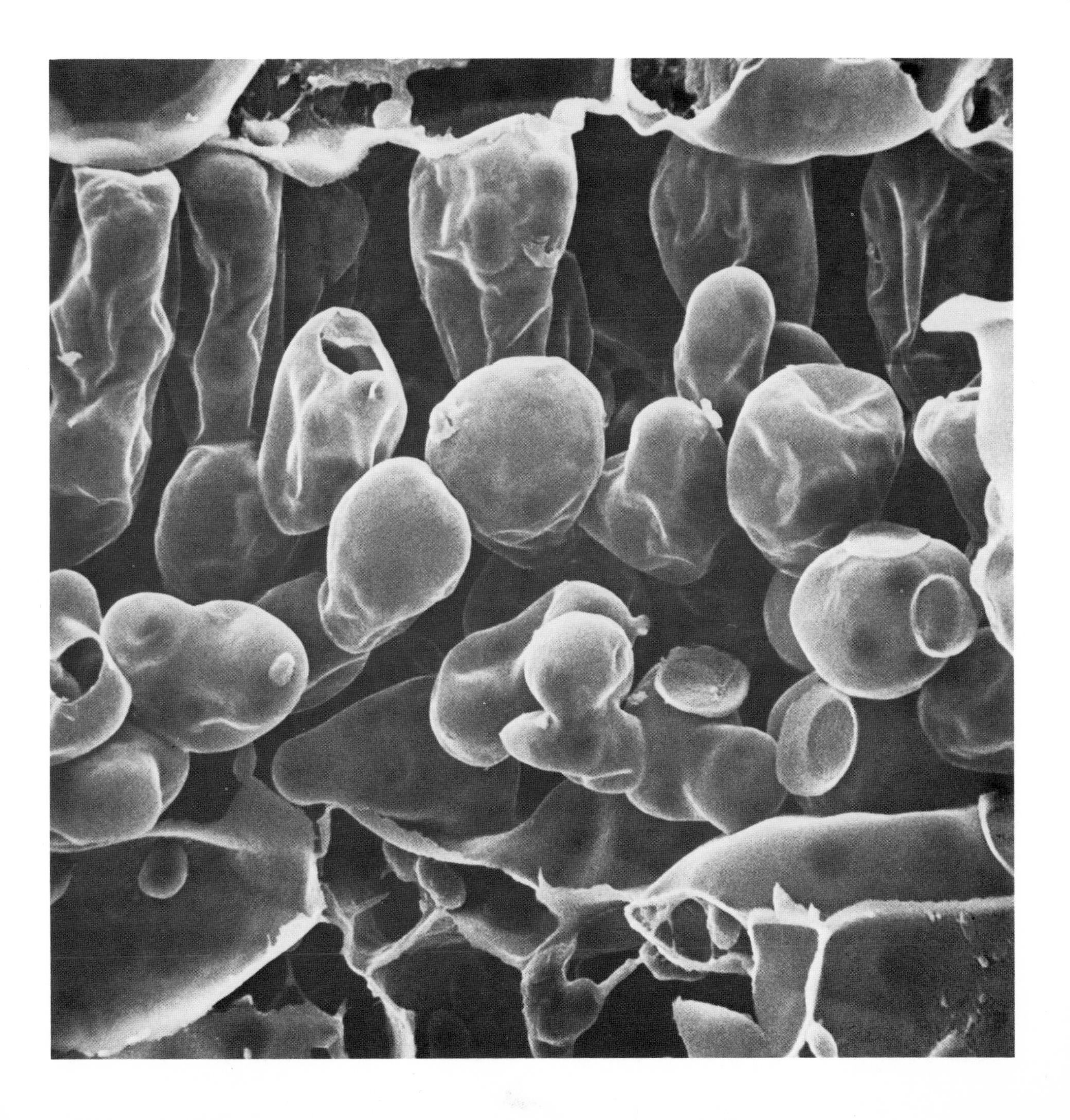

Plate 33

X 2,800

Peg-walled (tuberculate) rhizoids of *Marchantia*

There are two types of unicellular rhizoids on the lower surface. The smooth-walled type penetrate the soil and anchor the plant. The peg-walled type, shown below, which are of smaller diameter than the smooth-walled ones, lie parallel to the underside of the thallus. It has been demonstrated that their external walls form a capillary conducting system. They transmit water to the growing point, where it is readily absorbed. As the photo illustrates, the peg-like projections are on the internal surface of the wall. The significance of these wall ingrowths is unknown.

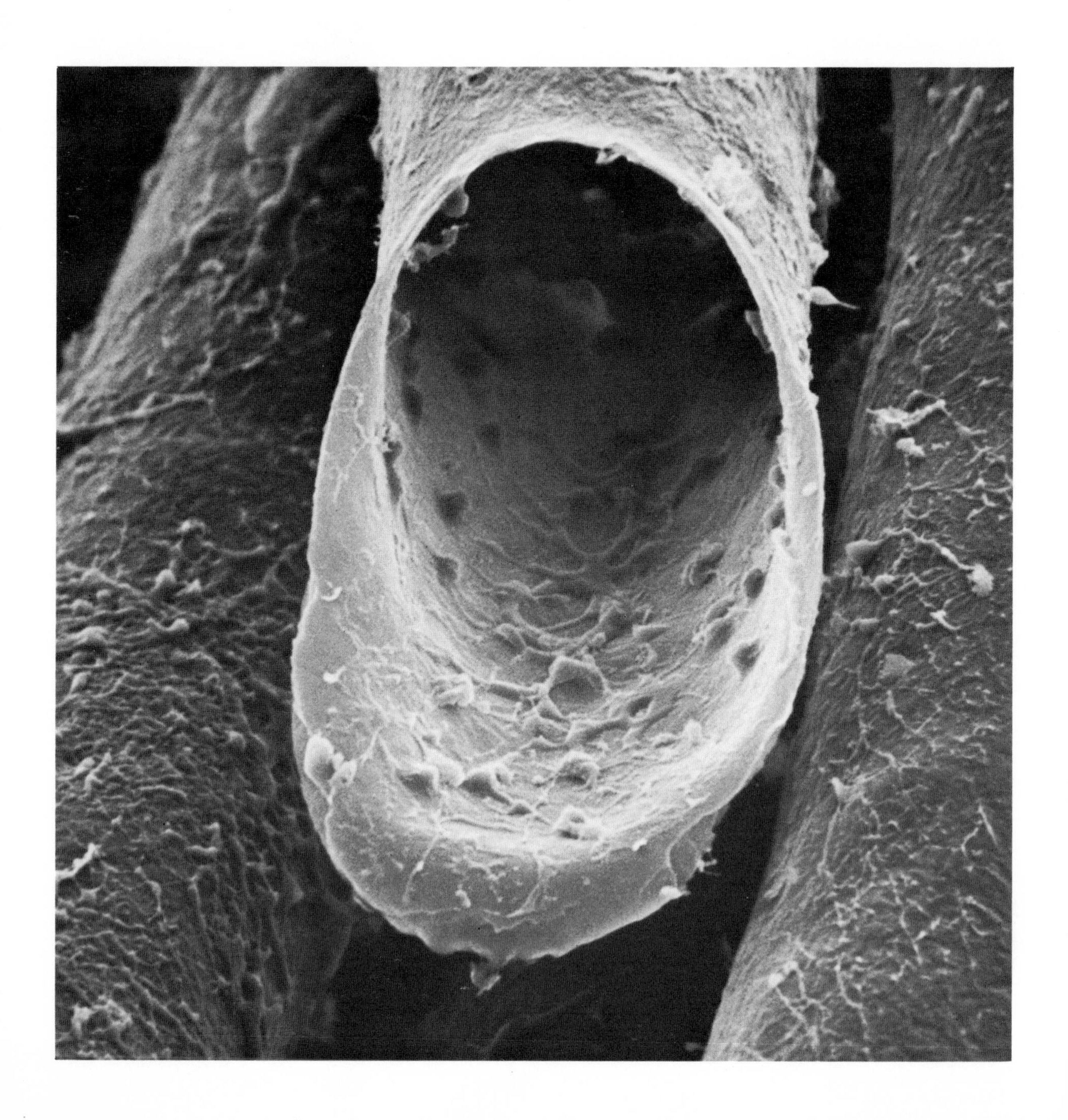

Gemma cup and gemmae of *Marchantia berteroana*

Plate 34
X 60

Gemma cups occur on the upper surface of the thallus. They contain gemmae—special asexual reproductive bodies. Each gemma is a multicellular, biconvex disc (Plate 35). Approximately halfway down the gemma are two lateral notches, opposite each other (**N**). Each gemma is attached to the floor of the cup by a large stalk cell and, when mature, the stalk disorganises (at (**C**)) and the gemma is dispersed by water.

When gemmae germinate on the soil, growth occurs in two directions from apical initials situated in the lateral notches. Most of the cells in a gemma contain chloroplasts, but there are some isolated colourless marginal cells from which rhizoids develop, if that particular face of the gemma lies against the soil. There are also scattered colourless cells containing oil bodies. Some barrel-shaped air pores can be seen on the outside of the cup (Plate 34).

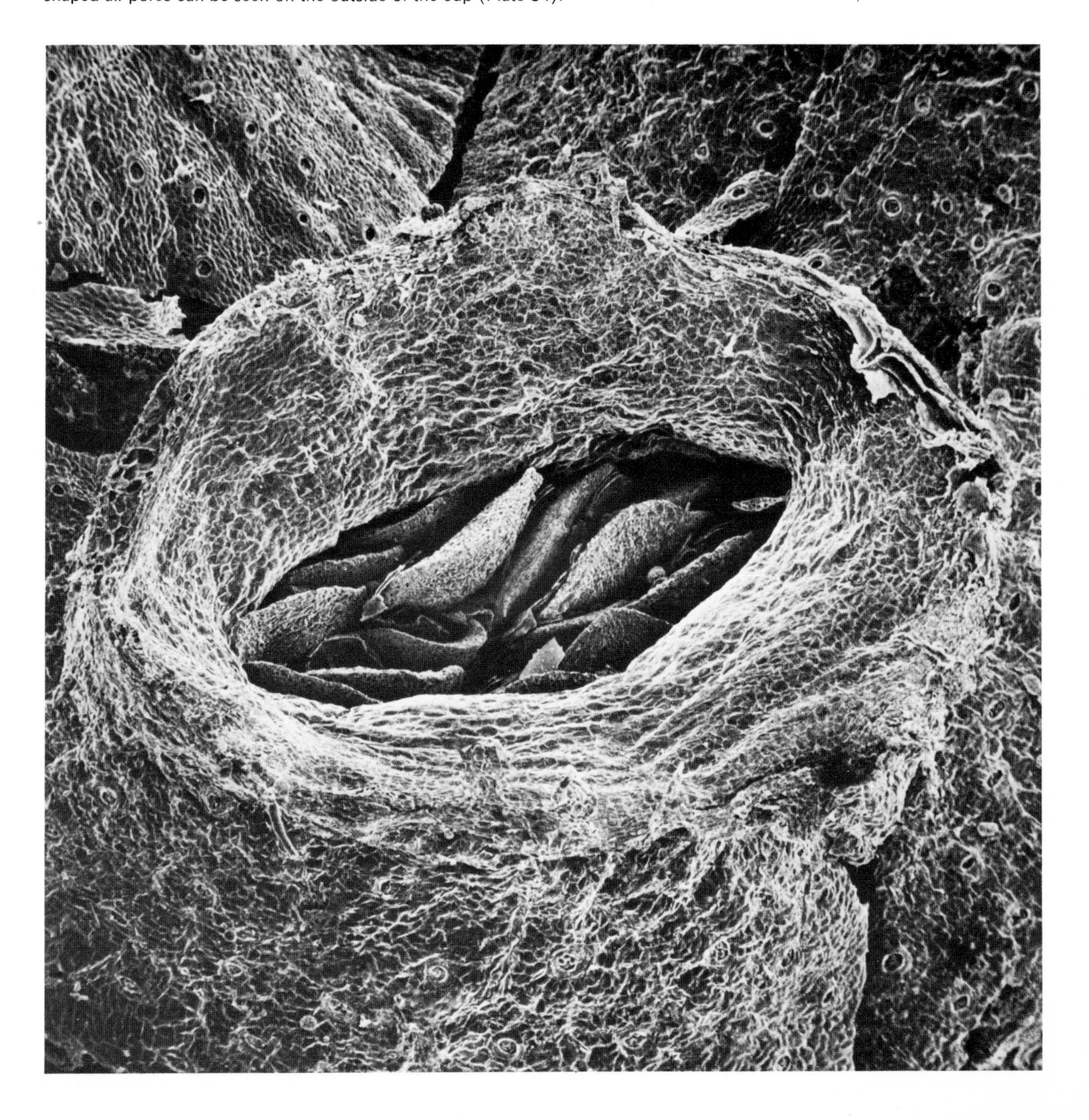

Plate 35
X 500

Gemma of *Marchantia berteroana*

See Plate 34.

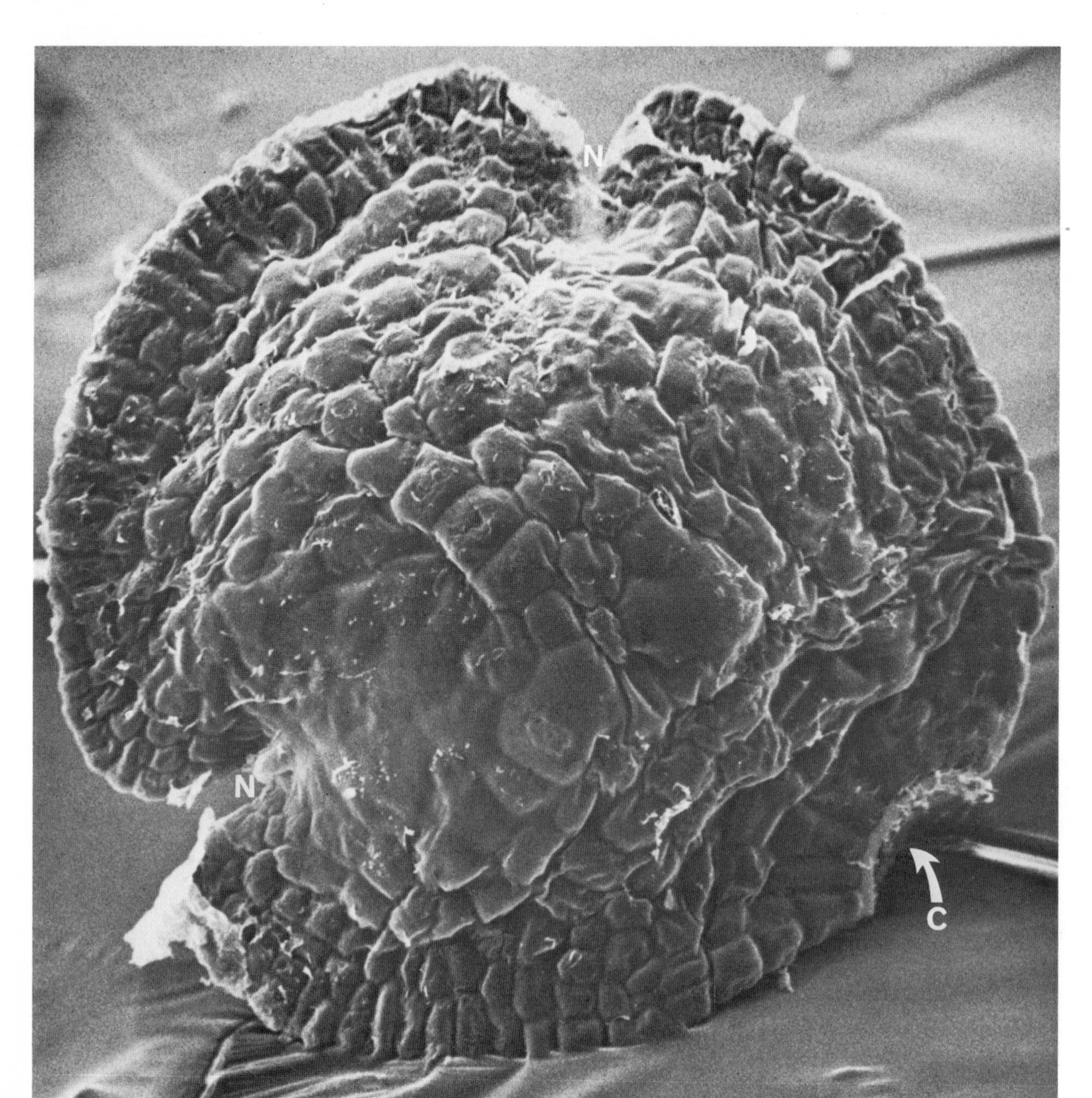

Spore and elaters from capsule of *Marchantia* sporophyte

Plate 36

X 2,000

Within the developing capsule wall of *Marchantia* there is, at first, only one type of cell, the sporogenous cells, which are vertically elongated. Eventually about half of them divide and redivide transversely, to form vertical files of spore mother cells. There are generally 32 spore mother cells in each file. Each spore mother cell undergoes meiosis to form four spores. The sporogenous cells which do not divide elongate further to become elaters. There is, then, a ratio of approximately 128 spores to one elater, although the photo does not give this impression because most spores have already been dispersed. Each elater has a double spiral of wall thickening. Under dry conditions the wall of the capsule splits at the apex into a number of segments to expose the spores and elaters. As the elaters dry they twist and shorten as the thinner parts of the walls shrink inwards and these jerking movements throw out spores. In some liverworts the elaters suddenly lengthen to their former size, but Ingold (1965) has noted that there is no such sudden and violent untwisting of elaters in *Marchantia*.

CLASS MUSCI

Mosses. The gametophytes are leafy with multicellular rhizoids. The sporophytes generally have a more complex dehiscence mechanism than liverworts.

Plate 37
X 400

Leaf surface of bog moss *(Sphagnum cristatum)*

There are over 300 species of *Sphagnum* growing in wet habitats throughout much of the world. The leaf is only one cell thick, without a midrib. Leaf structure is unlike that of any other plant. Large colourless water-filled cells with spiral wall thickenings alternate with smaller living green cells (Plate 37). The large cells, which have a caterpillar-like shape in this species, are dead at maturity.

Each large cell has many so-called pores which occur on both surfaces of each cell. They are aligned in two rows on each surface in this particular species. Each pore is inserted between the spiral thickening. The pore has a thickened rim (Plate 38). An SEM study by Mozingo, Klein, Zeevi and Lewis (1969) on *Sphagnum imbricatum* revealed that, as in *S. cristatum*, the pores are

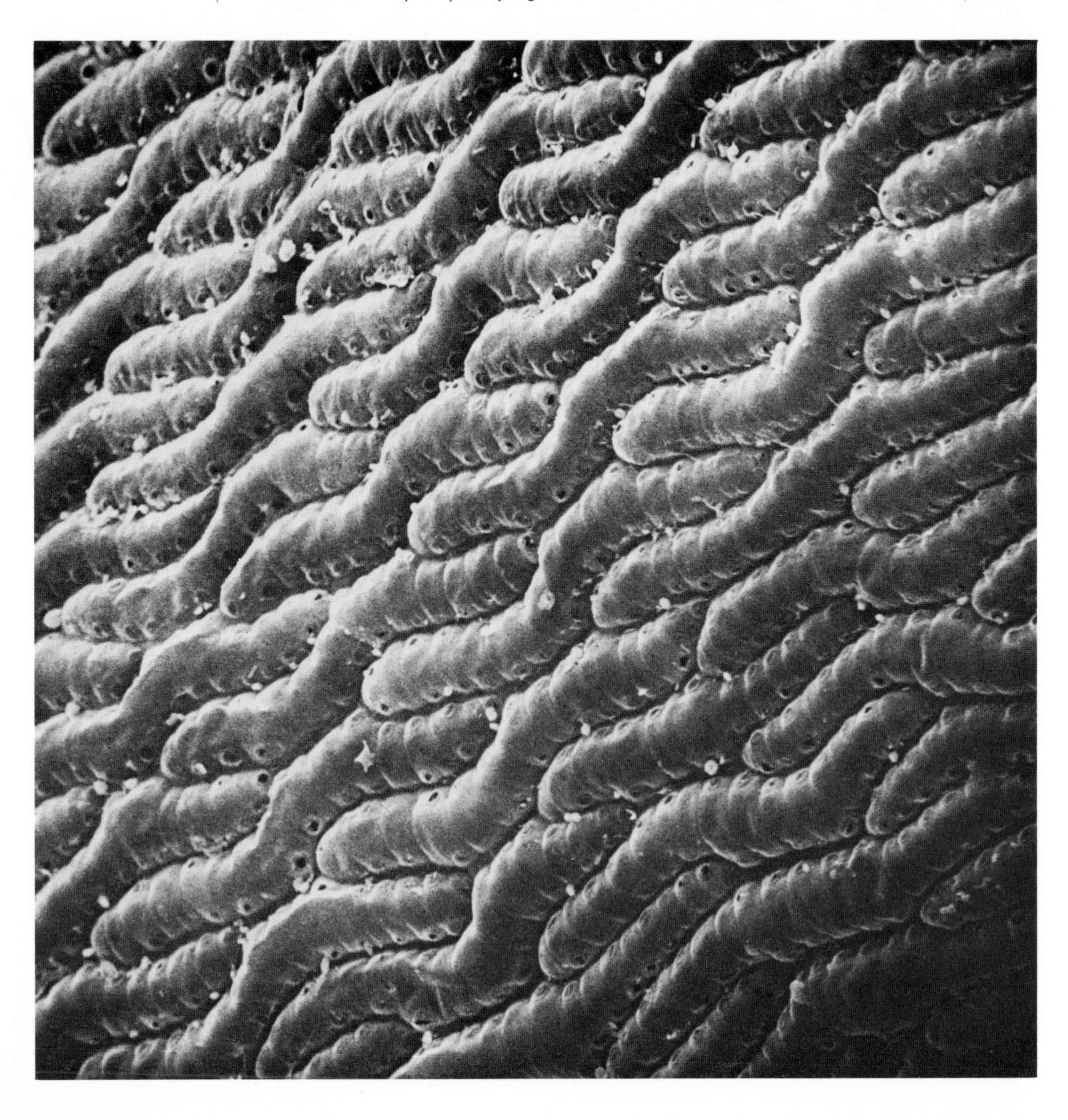

UNIVERSITY OF WINNIPEG
LIBRARY
515 Portage Avenue
Winnipeg, Manitoba R3B 2E9

covered by a membrane and some of the membranes rupture to form true pores. Two of these can be seen in Plate 38. It is not known whether membrane rupture is a natural phenomenon or the result of drying of the plant. The pores allow rapid absorption of water, while the spiral thickenings give mechanical support and prevent the cells from collapsing when empty. Each large cell is surrounded by a ring of the green living cells (**L**) which are so small that they occupy an area which looks like part of the walls of the large cells.

Because of its water-retaining properties, *Sphagnum* is used in horticulture as packing around young plants when they are transported. Peat moss can absorb and retain water weighing up to 20 times as much as the dry weight of the plant (Parihar, 1965).

Plate 38
X 2,000

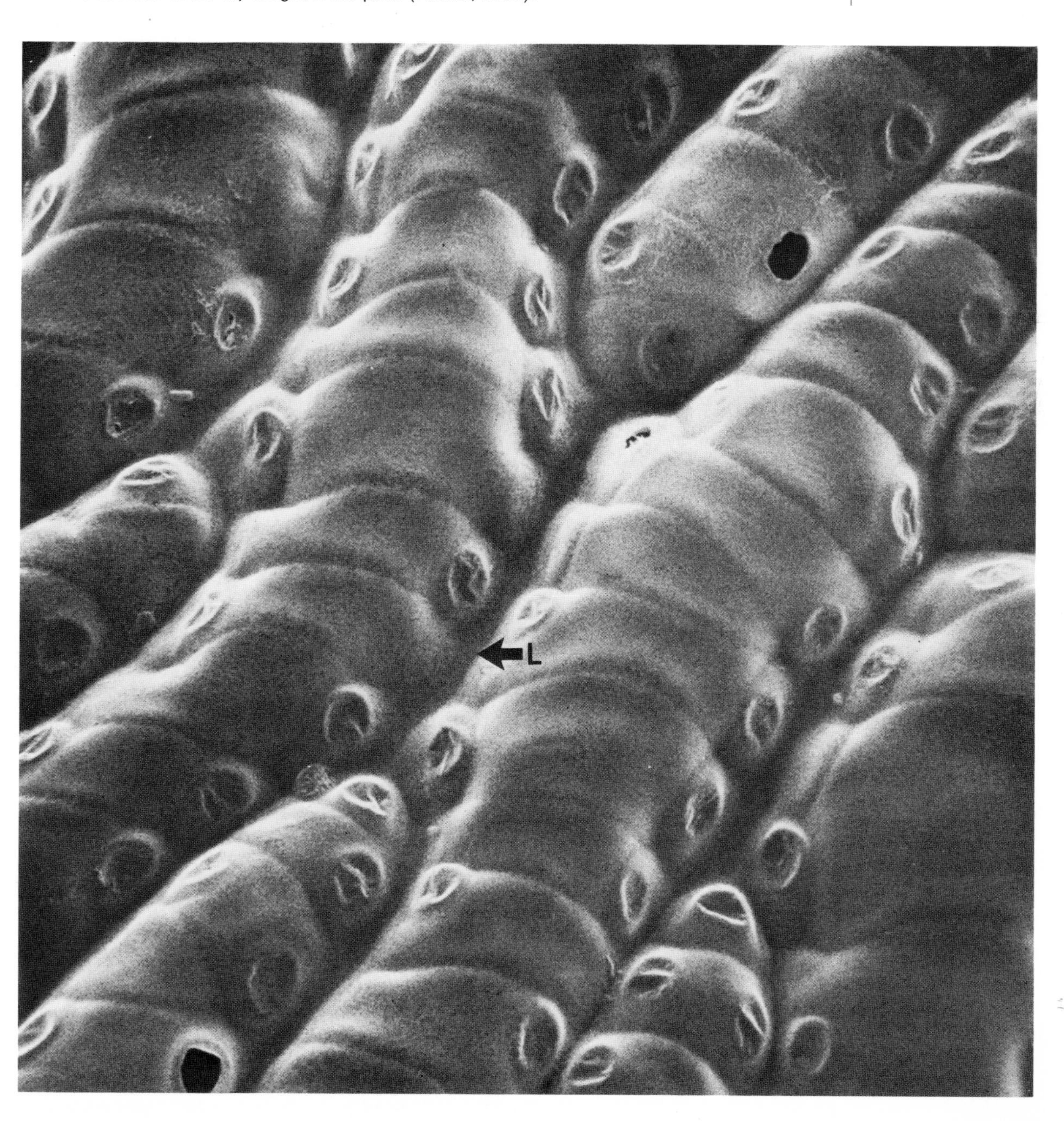

Plate 39

X 2,000

Transverse section of ***Funaria hygrometrica*** **Hedw. leaf**

The leaf of *Funaria* is only one cell thick, except in the midrib region, which is not shown in the photo. Chloroplasts can be seen within the cells and in the central cell a nucleus (**N**) is visible.

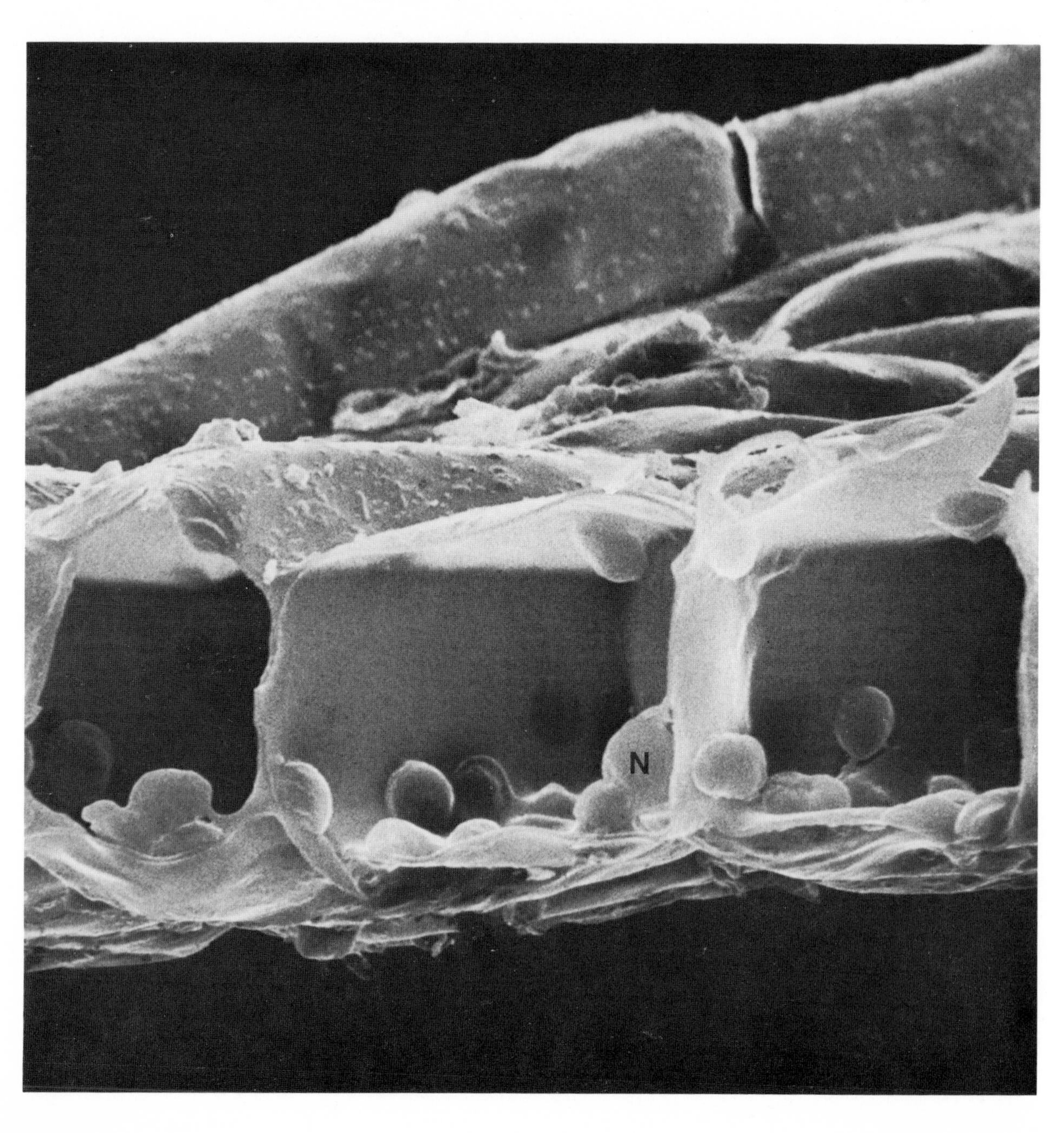

Upper leaf surface of the moss *Dawsonia superba* Grev.

Plate 40

X 900

This moss is unusual in having leaves many cells thick. The comparatively complex leaf anatomy is very similar to *Polytrichum*, which is illustrated in many texts. There is a large lower epidermis and internal to this are several layers of cells, including groups of small thick-walled ones. The upper part of the leaf comprises a layer of large cells from which arise plates of cells. These plates of photosynthetic tissue extend along the length of the upper surface. Each plate is one cell wide and up to six cells high. The plates are shown in surface view in Plate 40. Air is able to diffuse freely between the plates and such an arrangement of photosynthetic tissue is somewhat analogous to that of *Marchantia* (Plate 32).

Plate 41 shows a closer view of part of three plates. The surfaces are covered with flakes of wax, which conceal the boundaries between the cells along the surface of each plate. To our knowledge, this is the first report of surface wax deposits on the leaf of a moss. We observed similar wax deposits on leaves of *Polytrichum juniperinum* Hedw. The families Polytrichaceae and Dawsoniaceae have many features in common, including the same chromosome number of $n = 7$ (Watson, 1971).

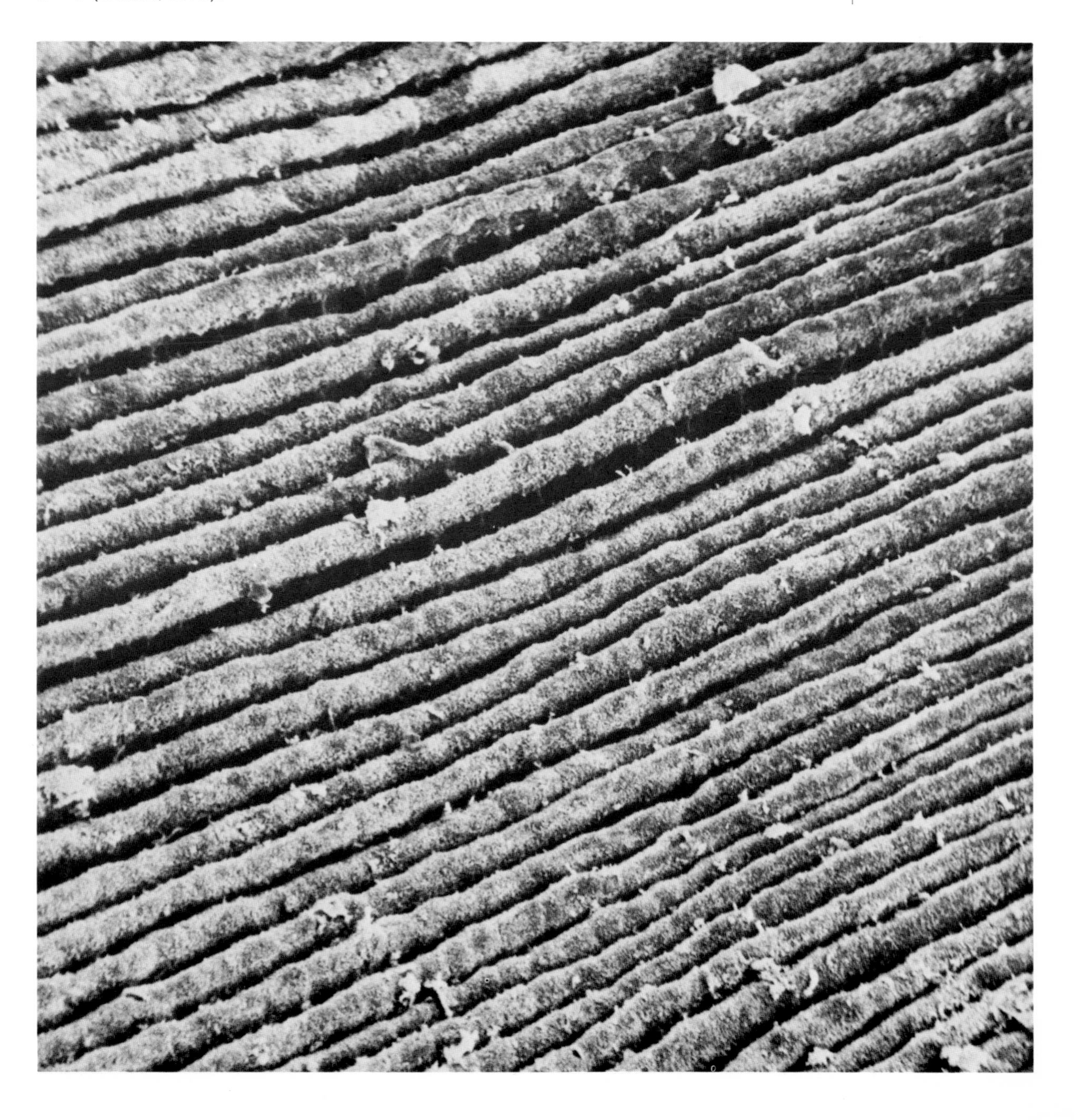

Plate 41
X 10,000

More magnified view of leaf surface of *Dawsonia superba*

See Plate 40.

Stomata on the capsule of *Funaria hygrometrica* Hedw.

Plate 42
X 500

Moss leaves do not have stomata. This is to be expected, as the leaves of many species are only one cell thick, except in the midrib region, and possess a very thin cuticle or lack one "in the ordinary sense" (Watson, 1957). Gases and water vapour can therefore readily exchange between the leaf cells and the atmosphere.

Stomata occur on the sporophytes of some mosses and can be seen in this view of the lower part of a *Funaria* capsule. Part of the capsule is shown in section, at lower left. Beneath the epidermis, which has a well-developed cuticle, there is extensive chlorophyllous tissue. When the sporophyte stalk elongates, the maturing capsule is elevated above the protection of the gametophyte leaves, which are close to the ground in a moist environment. This elevation of the capsule allows spore dispersal over a larger area, but the sporophyte is exposed to drier conditions. The cuticle and stomata provide protection against desiccation while the spores are maturing. The opening and closing of stomata control exchange of gases and water vapour between the cells of the capsule and the atmosphere.

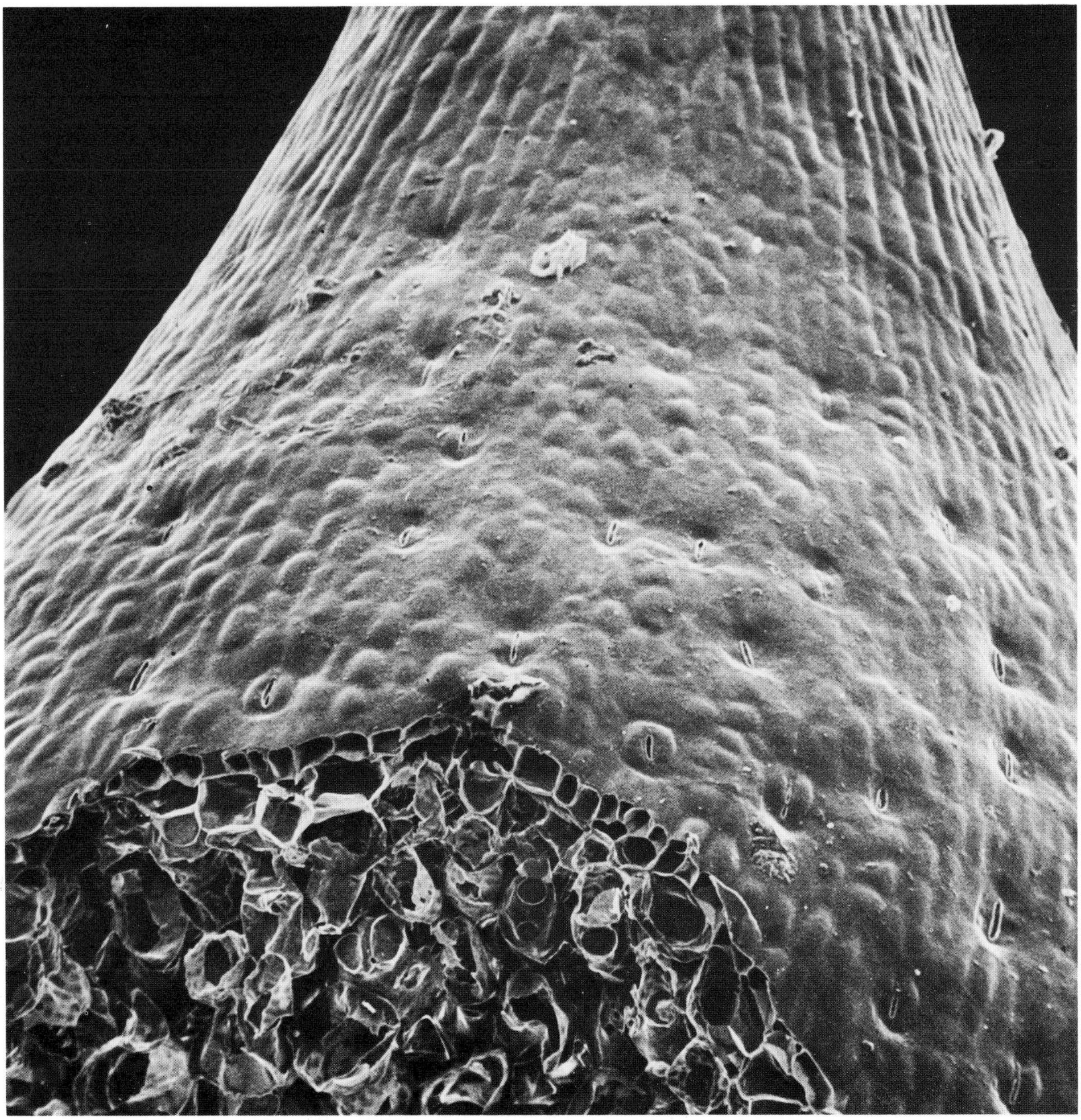

Plate 43
X 3,600

Stoma on *Funaria* capsule

A more magnified view of a stoma on the capsule illustrated in Plate 42. The cuticle covers the two sausage-shaped guard cells forming the stoma (pore) and conceals the boundary between these cells.

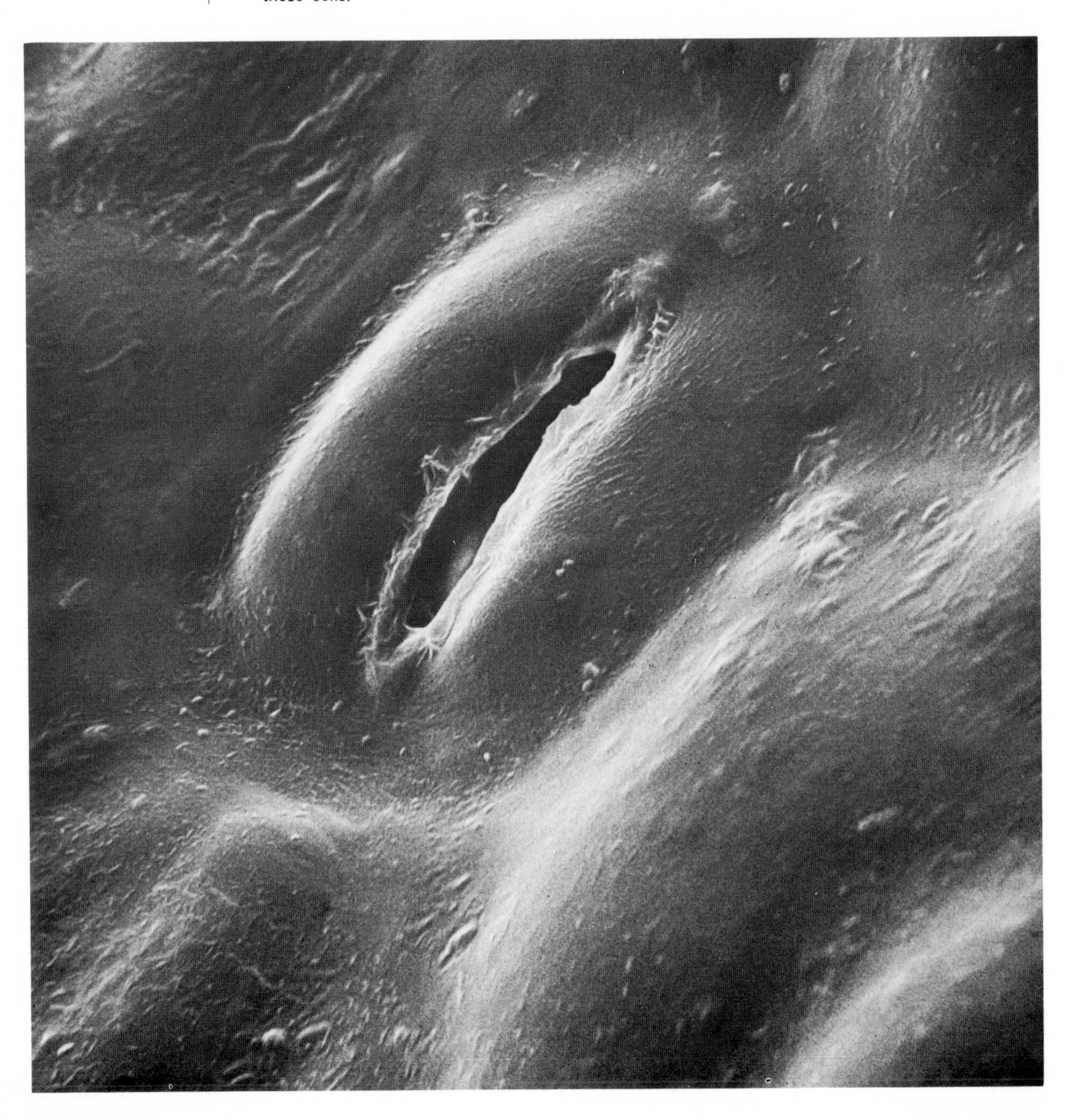

Transverse view of stoma on *Funaria* capsule

This transverse fracture of the basal part of a capsule passes through an open stoma. The cuticle can be clearly seen as an even layer which covers the outer surfaces of the external cells.

Plate 44
X 1,600

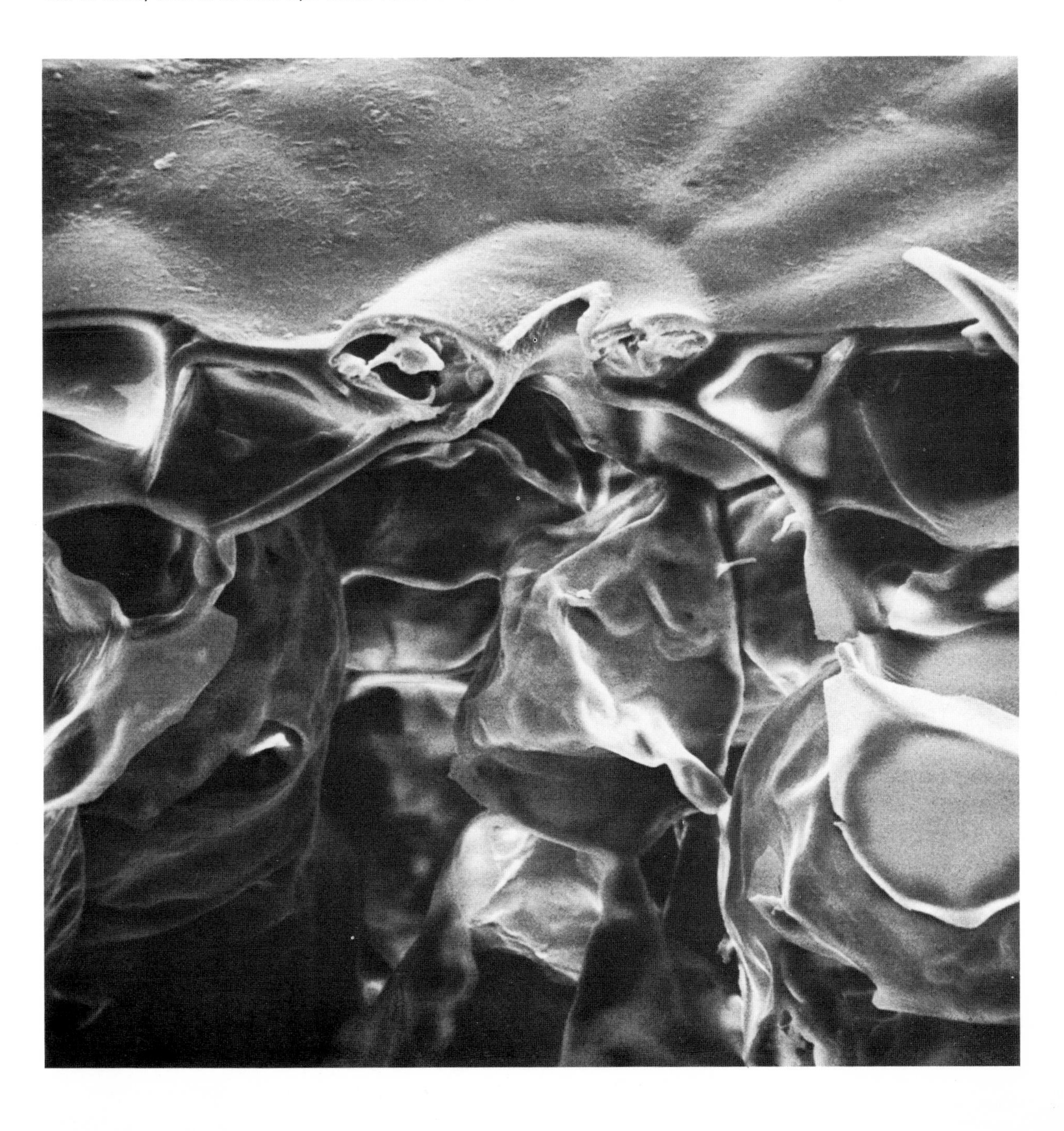

Plate 45

X 60

Calyptra on a young *Funaria* sporophyte

A new sporophyte generation begins when the egg in an archegonium is fertilised by a male gamete (spermatozoid) released from an antheridium, to form the zygote (one-celled embryo). For a time, the venter of the archegonium (now termed the calyptra) keeps pace in its growth with the developing embryo, thereby protecting it in its early development. The remains of the neck of the archegonium is arrowed. Later, the young sporophyte outstrips the growth of the calyptra which breaks transversely near the base of the venter and becomes carried upwards on the tip of the sporophyte capsule.

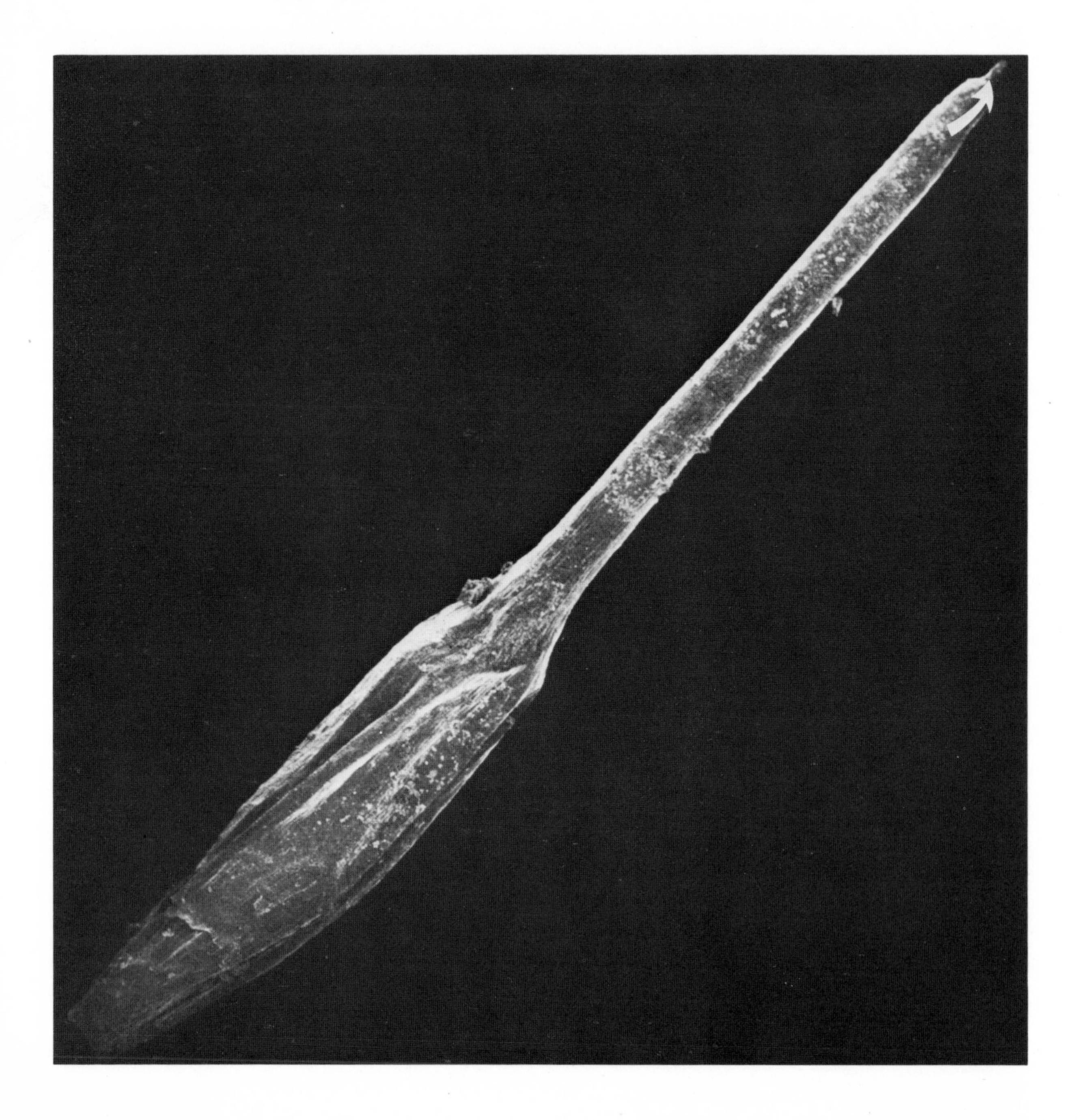

Tip of *Funaria* capsule showing operculum

Plate 46

X 250

The operculum is a cap at the top of the sporophyte. Like the peristome teeth which are below it, the operculum is derived from a dome-shaped cap of tissue, three cells thick (Proskauer, 1958). The outermost layer of the operculum is part of the sporophyte's epidermis. At the base of the operculum is the annulus (**A**) consisting of larger tissue, several cells high and three cells thick. The lowermost cells of the annulus are thin walled, forming an abscission zone which breaks down and the operculum is shed. It is thought that upward movements of the peristome teeth below assist in the shedding of the operculum. When the operculum is shed, the spores are still enclosed by the peristome teeth (Plate 48).

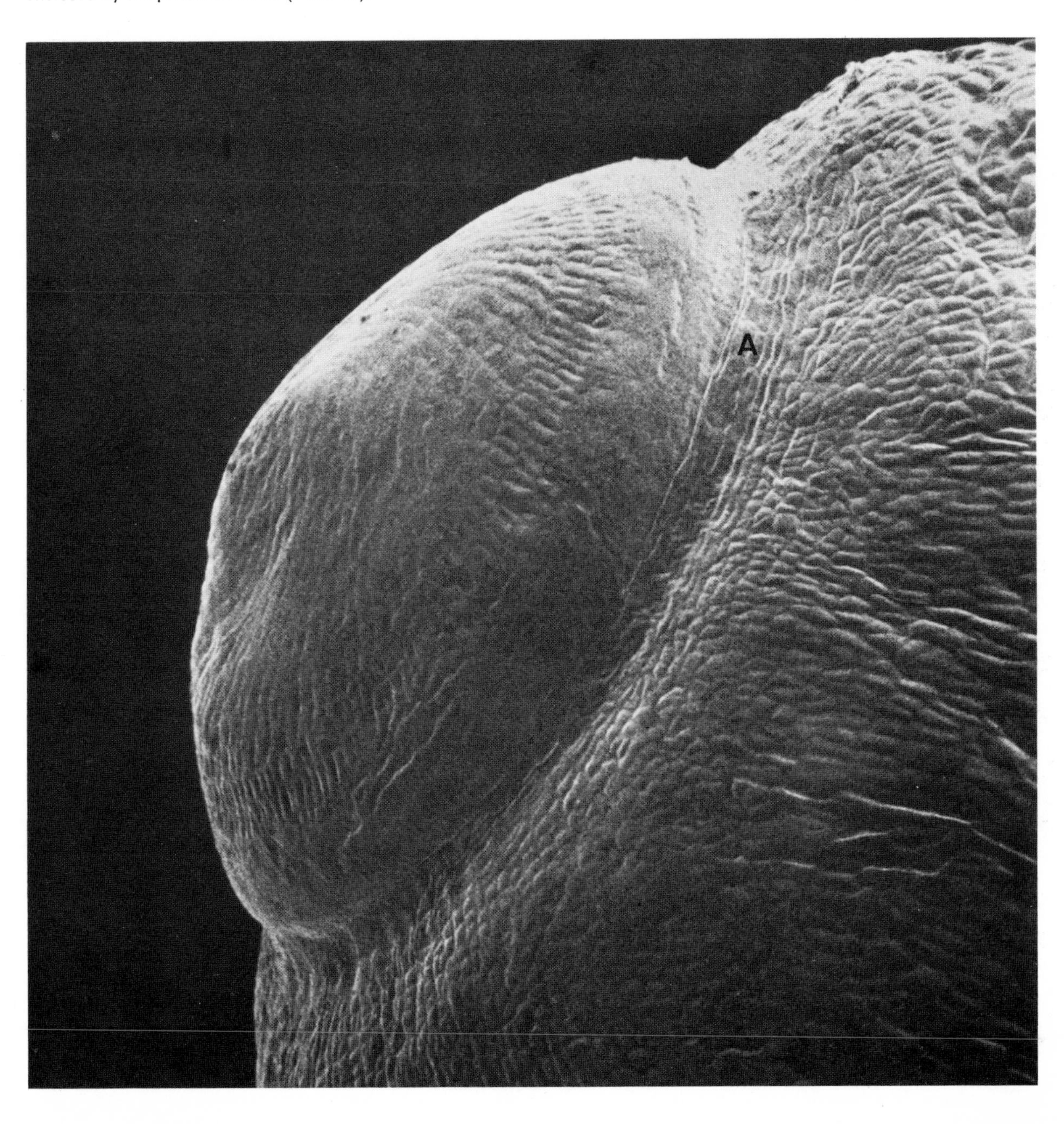

Plate 47
X 140

Side view of upper part of *Funaria* capsule after the operculum has been shed

The spore mass can be seen beneath the peristome teeth. A few of the inner ring of peristome teeth are partly visible.

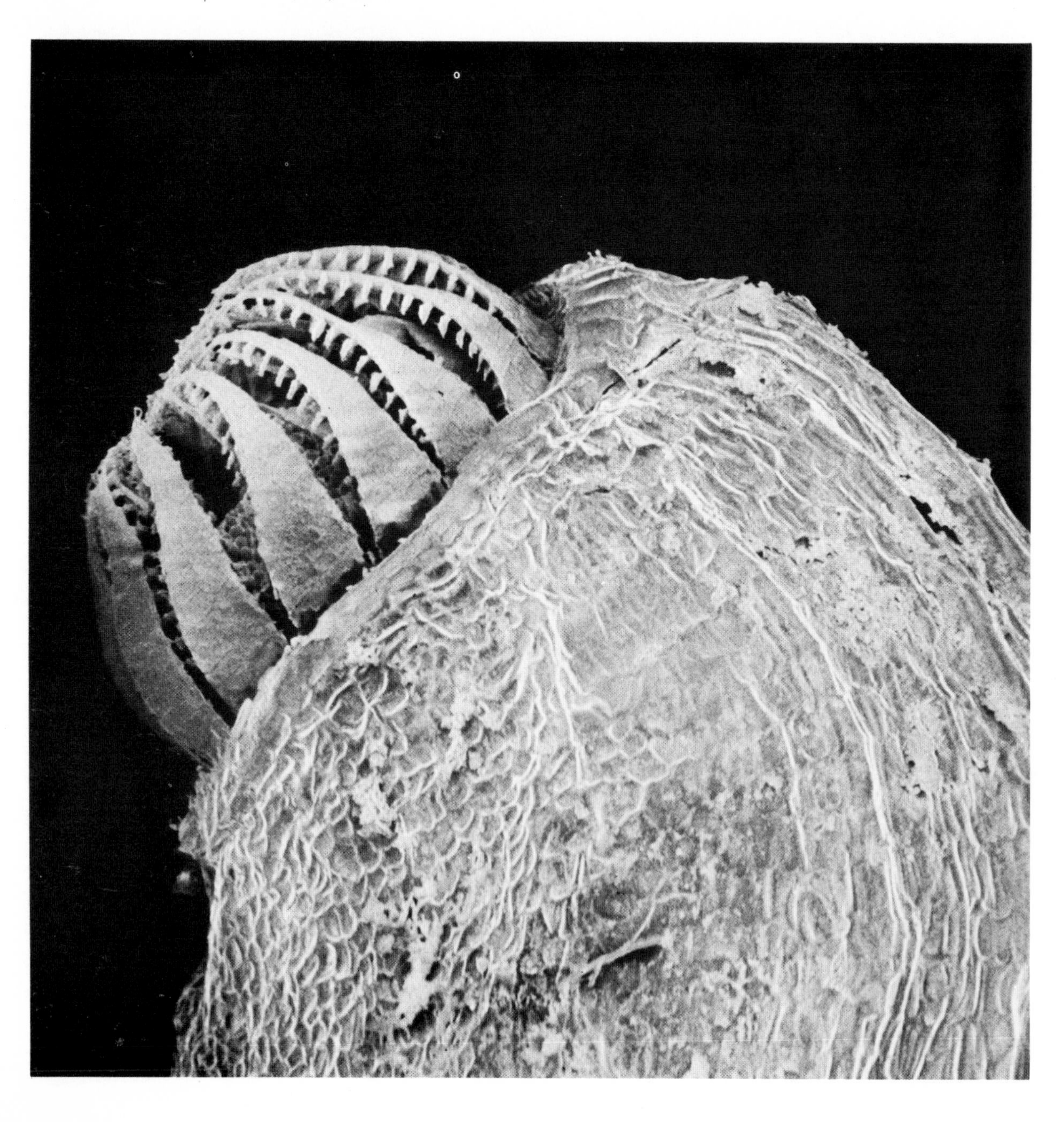

View of tip of *Funaria* capsule without operculum

Plate 48

X 150

At this stage the tips of the outer ring of 16 teeth have not separated from a central grid of cells (**G**), but the teeth are far enough apart to enable spores to begin to disperse from the capsule. The inner ring of teeth is not visible. Proskauer (1958) has noted that despite the number of investigations that have been made on capsule development in *Funaria*, this grid connecting the tips of the outer teeth has not received adequate study. In most specimens of *Funaria hygrometrica* that were examined, the tips of the outer teeth do not break away from the central grid before the spores are dispersed and even in old wrinkled capsules the tips of the teeth are still not free. This feature, which is typical of *F. hygrometrica* (Ingold, 1965) does not apply to most mosses with peristome teeth. We did find a few specimens of *F. hygrometrica* in which the outer teeth had parted from the grid.

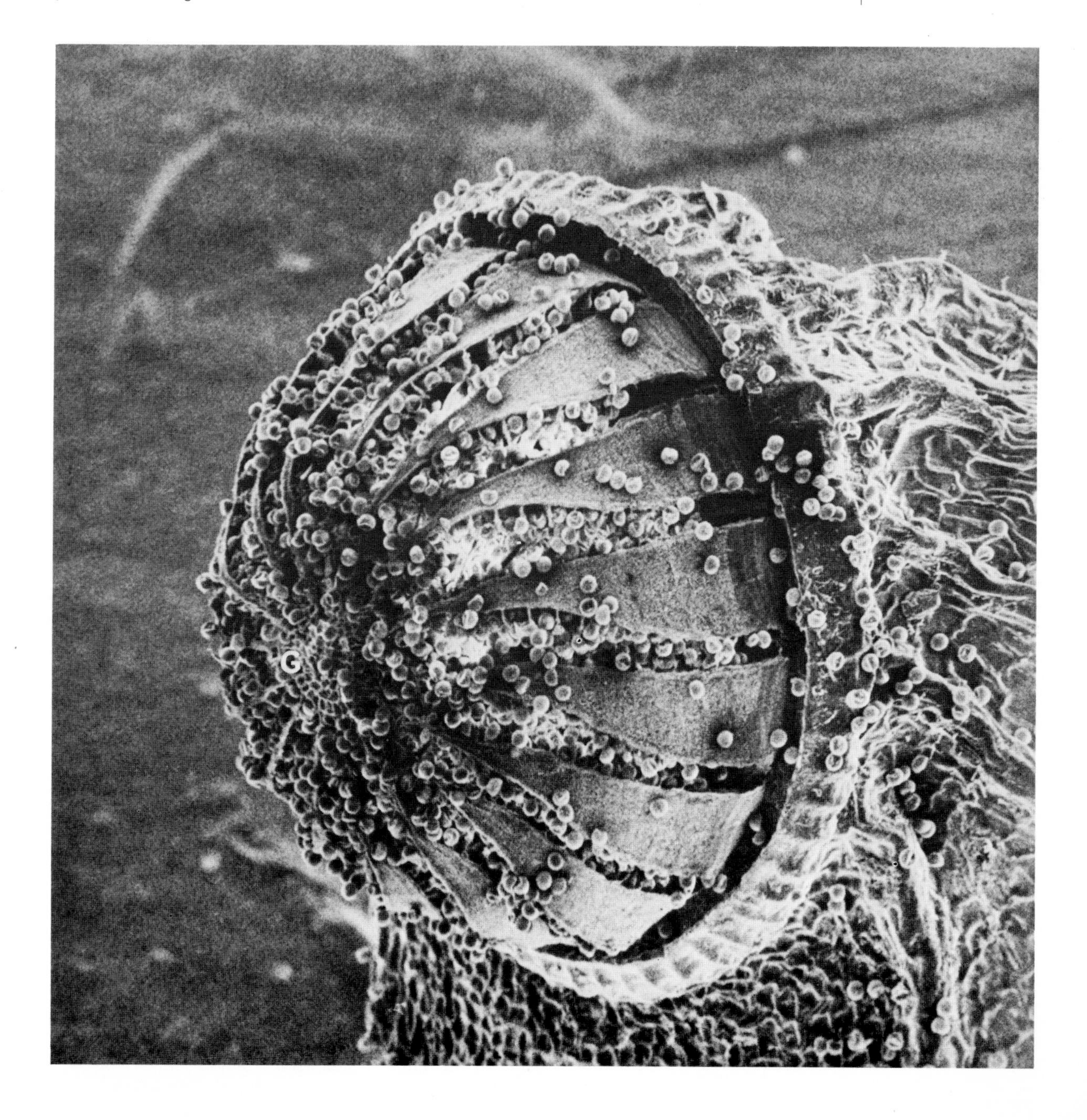

Plate 49

X 280

Fully open capsule of *Funaria hygrometrica*

This photo reflects an artificial situation in that, under natural conditions, the teeth do not bend outwards to this extent. The 16 teeth of the outer ring bend outwards or inwards with changes in humidity but usually remain attached by their tips to a central grid of cells (Plate 48). The two outer teeth at the top of the photo bear at their tips some of the remnants of this grid. The teeth consist of old cell-wall material and each tooth is a two-ply structure. Under damp conditions the outer layer absorbs more moisture and increases in length to a greater extent than the inner layer and the tooth curves inwards. On drying, differential shrinkage causes the tooth to bend outwards. Consequently the spores are released during dry weather. The tips of the smaller and more delicate teeth of the inner ring protrude outwards and readily catch against the jagged transverse bars on the outer teeth, when the latter are bent inwards. An inner tooth can be caught against an outer tooth and dragged outwards placing it under strain. It may then slip free, but often in a series of jerks as it becomes caught against successive bars on the outer tooth. During these movements and when the inner tooth is finally free, spores adhering to the tooth are flicked from the capsule.

Funaria is unusual among mosses in having the inner teeth on the same radii as the outer, rather than alternating with them. The structure of the peristome is a useful taxonomic character as there is considerable variation among different groups of mosses. Some possess a single ring of teeth and in a number of groups there has been independent evolutionary reduction in the peristome. Thus, some species of *Funaria* have either a rudimentary peristome or have completely lost it.

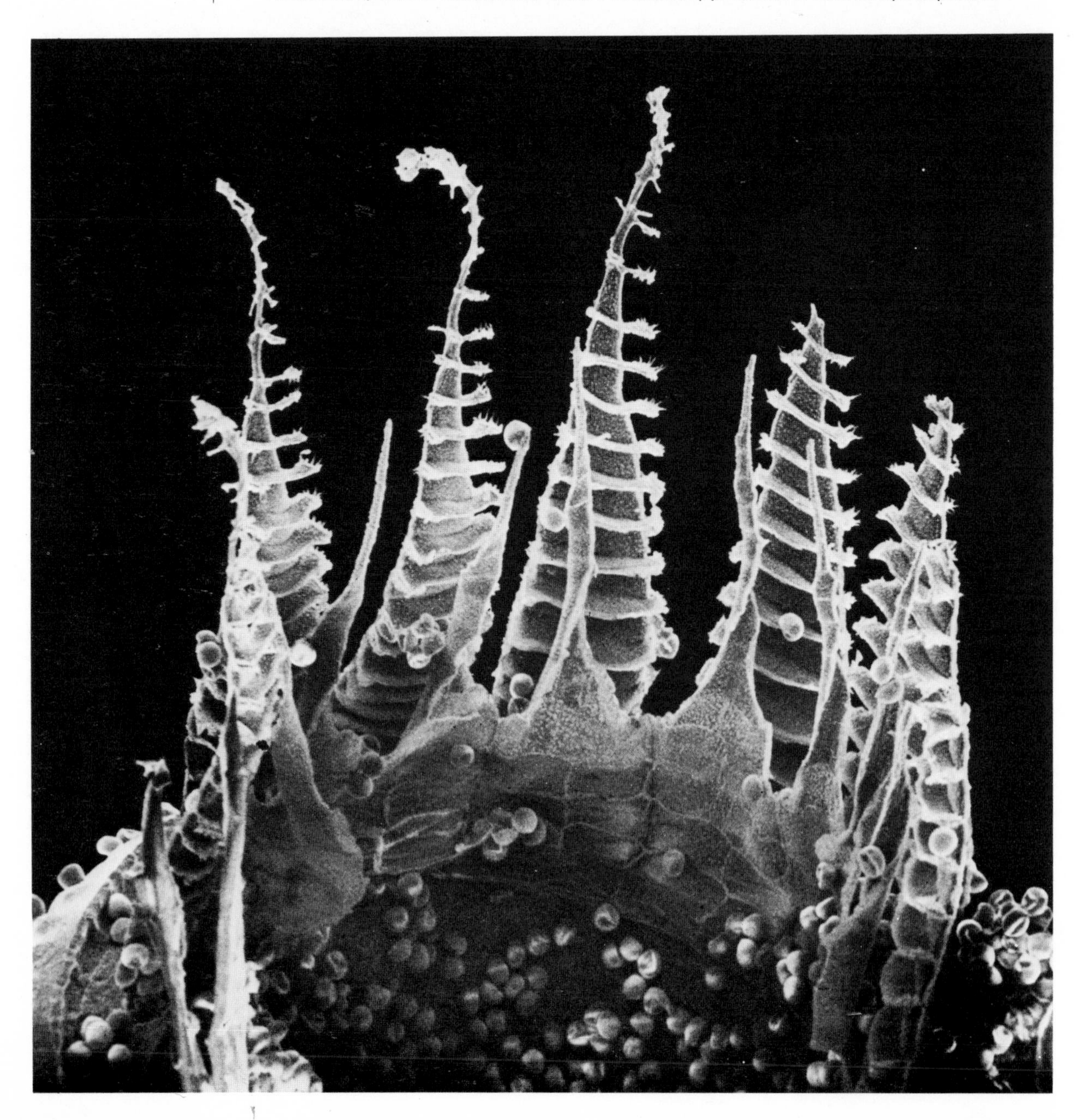

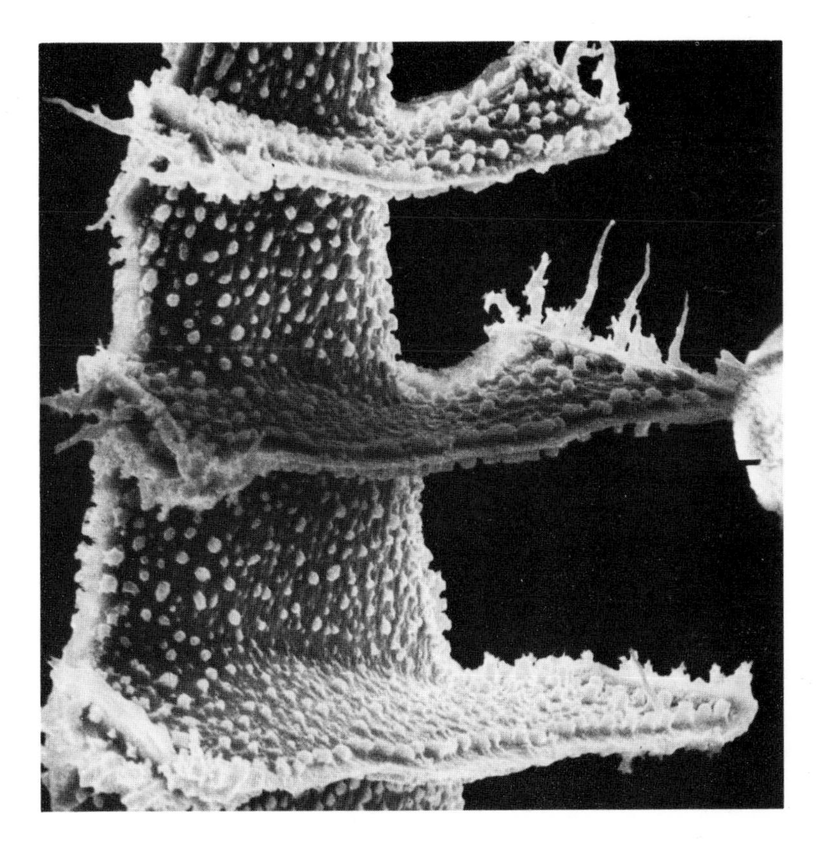

Plate 50 X 1,700

Part of the inner surface of an outer peristome tooth of
Funaria

Formation of the peristome teeth is an intricate process. They are derived from a dome-shaped cap of tissue, three cells thick, just beneath the operculum. The outer and inner layers are 32 cells in circumference and the central one is 16. In a transverse section of the dome near its apex, there would therefore be a ring of 32 cells in the outer layer, but these cells would of course be much smaller than those in a section near the base of the dome. The papillose thickenings on the teeth, which can be seen in the photo, are first laid down on the inner tangential walls of the outer layer of cells and the outer tangential walls of the inner layer. At a later stage, this thickening is deposited on the inner and outer walls of the middle layer. These outer walls are first coated by a more homogeneous type of thickening (Proskauer, 1958).

The three layers split in such a way that all cells are destroyed to form two caps of cellular remnants, one on top of the other. The outer cap, which forms the outer ring of teeth, develops from the junction of the outer and middle layers and the inner cap develops by splitting at the junction of the middle and inner layers.

The individual teeth are formed when there is vertical splitting along the unthickened anticlinal walls (those at right angles to the surface), which run from the apex to the base of the dome. Sixteen rather than 32 teeth are formed in each ring, because there are only 16 cells around the circumference of the middle layer. The three transverse bars in the photo represent parts of the anticlinal walls of the middle layer, which are at right angles to those anticlinal walls which split to form the teeth. It can be seen that the bars also have the papillose thickening. In this description of teeth formation, we have, for simplicity, ignored the central grid of cells which connect the tips of the outer teeth (Plate 48).

Plate 51 X 4,400

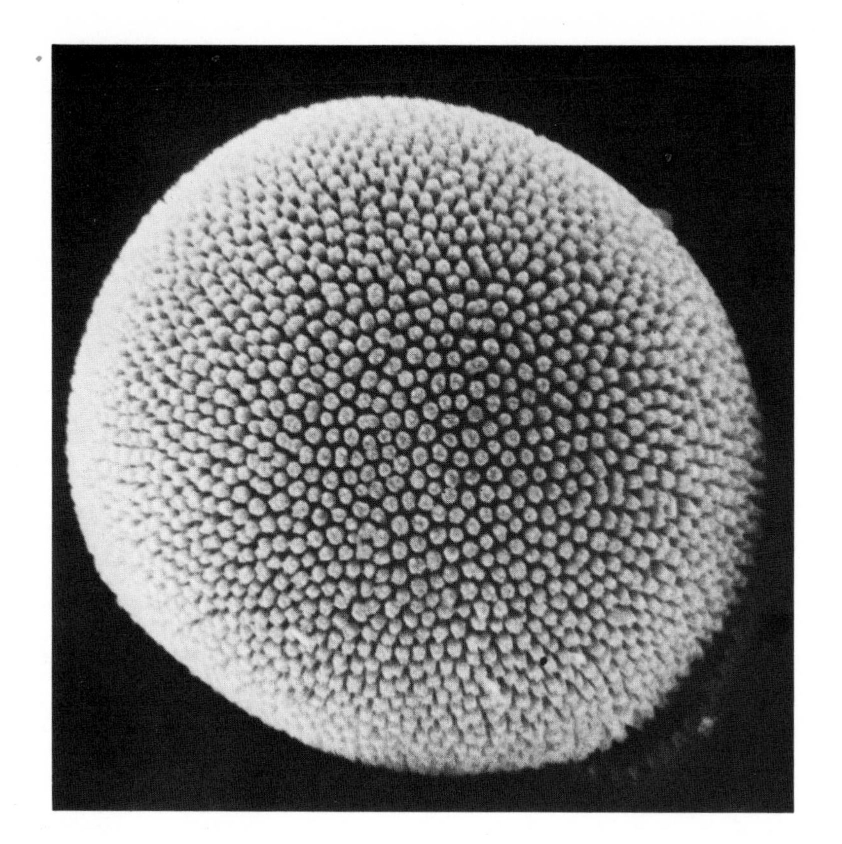

Spore of *Funaria hygrometrica*

This is a mature spore at the stage when it is shed from the capsule. *Funaria hygrometrica* is the best known and most widely distributed of all mosses and the spores are usually described as having smooth or almost smooth coats. However the papillose nature of the outer-wall layer (outer exine) shown in this photo can be seen at high magnifications under the light microscope. Internal details of spore structure and the nature of the wall, as seen with the transmission electron microscope, can be found in Monroe's (1968) paper.

Spores are produced by the meiotic division of spore mother cells in the sporophyte capsule and they therefore mark the beginning of the new gametophyte generation. The spores may remain viable for several years and on germinating they form a protonema. This branched filamentous structure develops buds which grow into the leafy moss gametophores. Antheridia or archegonia occur at the tips of the gametophores.

Antheridia of *Dawsonia superba* Grev.

Plate 52
X 170

We were unable to obtain good preparations of *Funaria* antheridia and used this Australasian moss which is one of the largest known, with gametophores up to 70 cm high. A number of antheridia, at various stages of development, are borne at the apex of male gametophores, among the terminal leaves. Leaves of the antheridial cup differ from the other leaves in being shorter, with wider bases, and have a short point in place of the normal leaf blade. The long cylindrical antheridia taper towards their rounded tips and have short multicellular stalks which are not visible in the photo. The sterile wall (jacket) of the antheridium is one cell thick and encloses the male gametes. Dehiscence of mature antheridia occurs when water has collected in the cup formed by the terminal leaves. Larger cells at the tip of the jacket absorb water and break up and the mass of biciliate motile spermatozoids is released from the top of the antheridium.

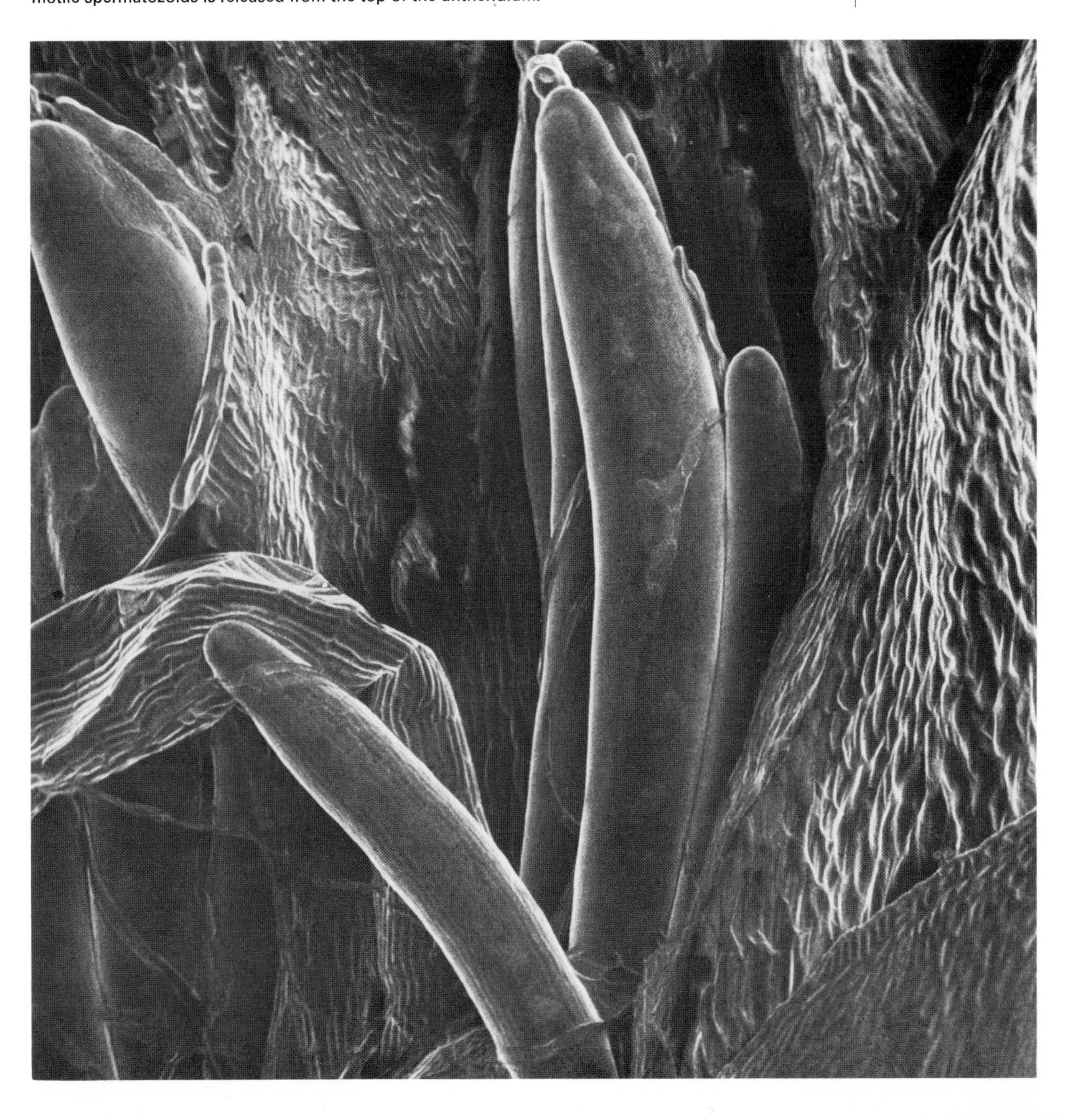

Plate 53

X 250

Archegonial group in a moss (*Cyathophorum bulbosum* (Hedw.) C.M.)

In this particular moss the sex organs are borne terminally on small branches in the axils of two rows of large leaves which run down the stem of the gametophore. Small leaves on each branch, which enclose the archegonial group, have been partly removed to expose the archegonia. Intermingled with the archegonia are sterile hairs, the paraphyses. Each paraphysis consists of a uniseriate filament of cells.

Each archegonium has a long neck (**N**), an expanded venter (**V**) which encloses the ovum and a thick stalk (**S**) which merges with the venter.

PHYLUM TRACHEOPHYTA

The so-called "vascular plants". They usually have well-developed roots, leaves and stems and conducting tissue is differentiated into xylem and phloem. The sporophyte generation is the conspicuous one.

SUBPHYLUM PSILOPHYTINA

Comprises two living genera, *Psilotum* and *Tmesipteris*. The sporophytes are dichotomously branched and have rhizomes but no roots. The sporangia are fused in twos or threes. The gametophytes are subterranean and do not possess chlorophyll.

Sporangia of *Psilotum nudum* (L.) Beauv.

Plate 54

X 80

Psilotum and *Tmesipteris* have been regarded as the living representatives of an ancient group of plants, the Psilophytina, the other members of which became extinct in the Palaeozoic era, more than 300 million years ago. Recent research, however, suggests that the affinities of *Psilotum* and *Tmesipteris* are with certain primitive ferns (Bierhorst, 1971). *Psilotum* is widespread in the tropics and subtropics and reaches as far north as Florida and as far south as New Zealand.

The small green aerial stems of *P. nudum* grow to three feet high. They bear small scale-like appendages often described as "leaves" although they lack veins. The sporangia usually occur fused together in threes. They occur terminally on what has been interpreted as a very short lateral branch which bears two "leaves", one of which (**L**) can be seen below. The sporangia have opened to release most of the spores.

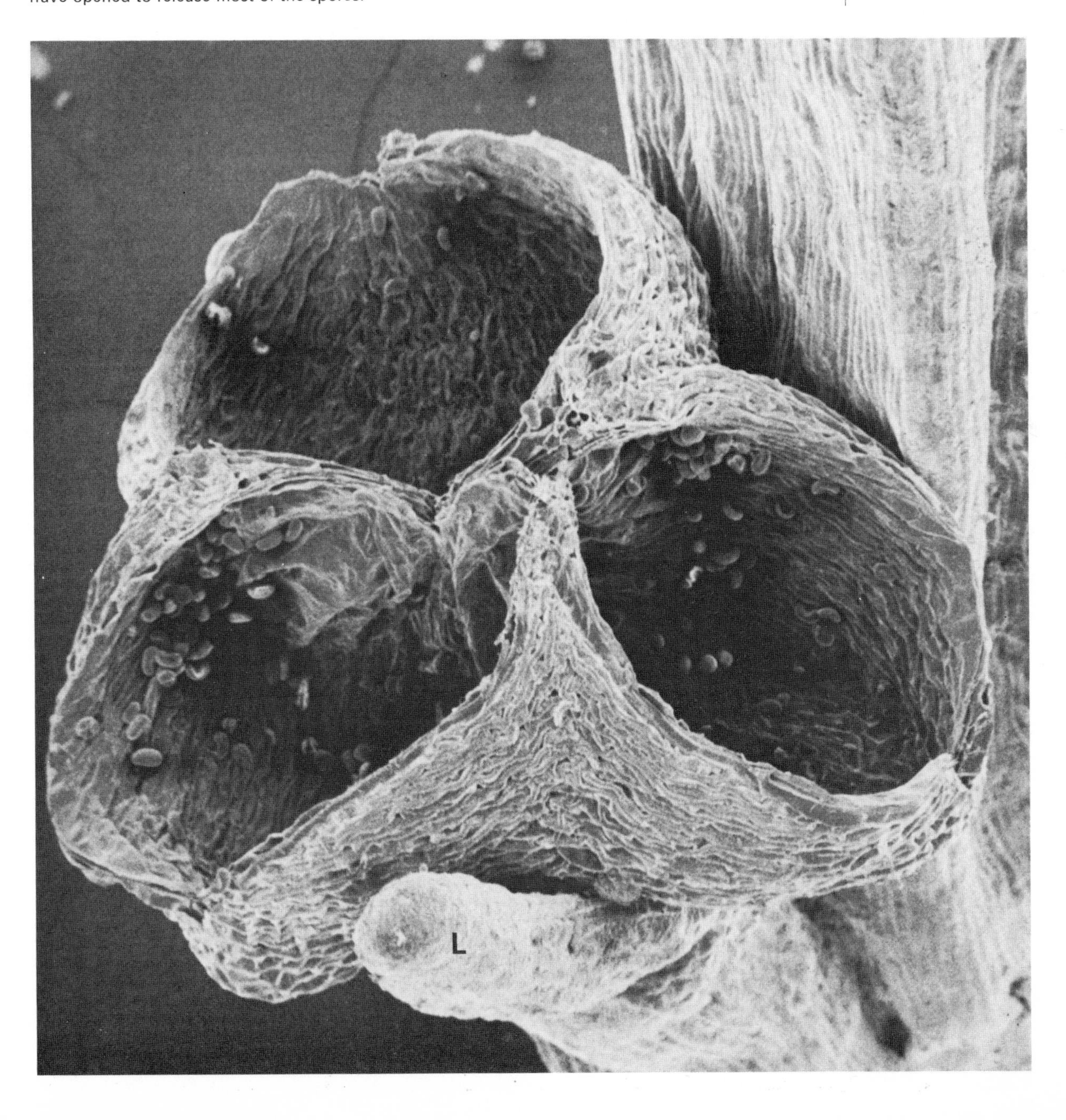

Plate 55
X 4,000

Stoma on stem of *Psilotum nudum*

The stomata occur in shallow longitudinal grooves running down the stem. They are arranged parallel to one another as in most monocotyledonous angiosperms (Plate 120). As in *Tmesipteris*, the guard cells are sunken in shallow pits. The outer walls of the guard cells in *Psilotum* and *Tmesipteris* have intricately sculptured surfaces which are presumably wax deposits. There are no stomata on the "leaves".

Both genera have the perigenous type of stomatal development. In this type, found also in grasses (Plate 125), the guard mother cell and neighbouring cells are not formed from the same parent cell. *Equisetum* (Plate 62) is an example of the contrasting mesogenous type of development. Epidermal cells of *Tmesipteris*, excluding guard cells, have surface walls of a unique type with lignified bands alternating with unlignified areas (Pant and Khare, 1971).

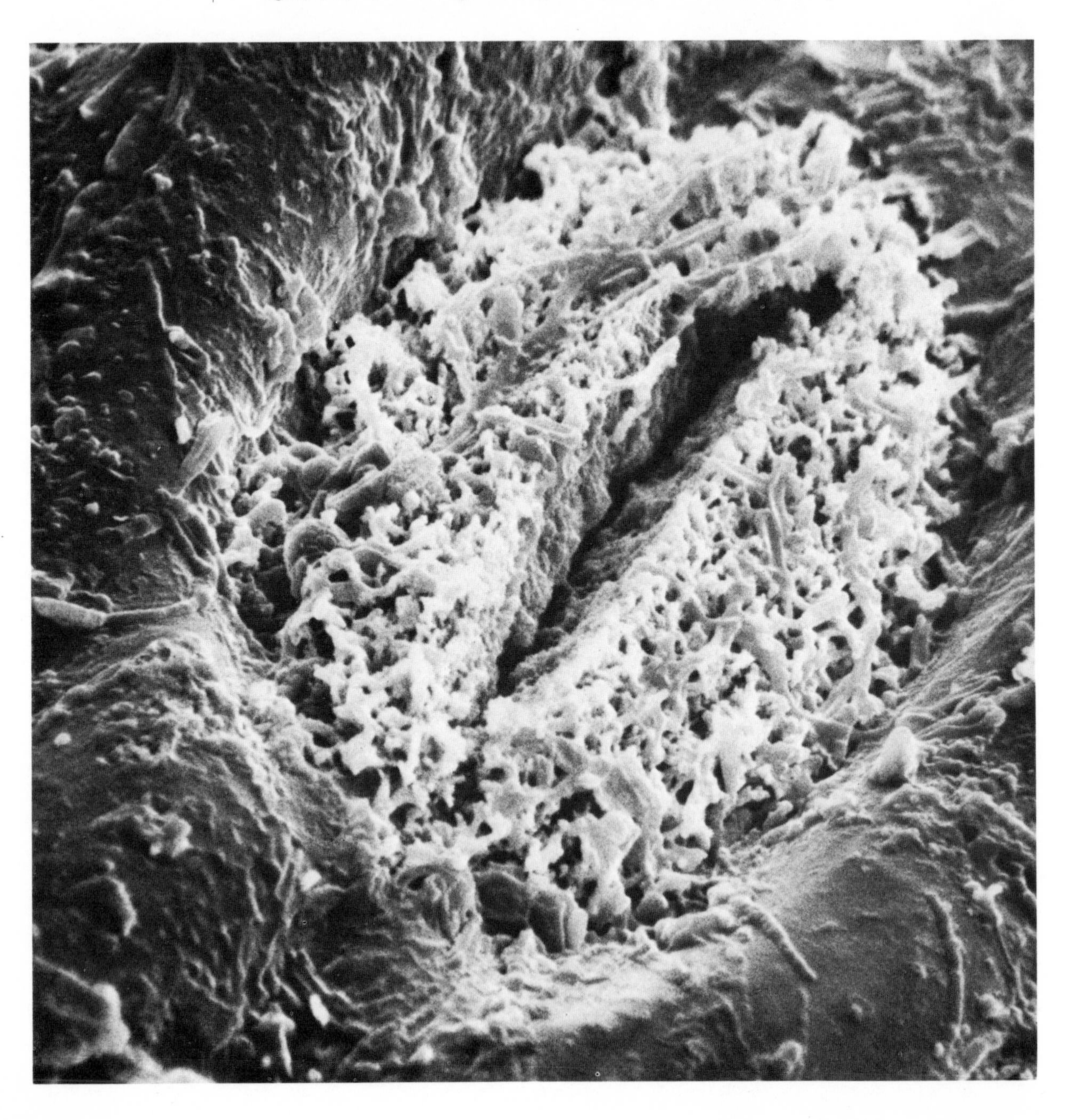

Sporangia of *Tmesipteris tannensis* Bernh.

Plate 56
X 50

Tmesipteris occurs in Australasia, Polynesia, New Caledonia and parts of Asia. Most species are epiphytes and stems grow to about a foot in length. The "leaves" are up to 2 cm long. The sporangia, as in *Psilotum*, are found on what has been interpreted as very short lateral branches, each bearing two leaves and terminating in, normally two, fused sporangia. The sporangium at right has begun to split along the line of dehiscence.

Plate 57

X 840

Spores of *Tmesipteris tannensis*

Psilotum and *Tmesipteris* have very similar spores. Each is a bean shape with a slit-shaped depression on the concave side. The spore germinates into a gametophyte (prothallus) which develops beneath the soil and bears archegonia and antheridia. These subterranean gametophytes of both genera are very similar. They are colourless, sparingly branched, cylindrical structures which bear rhizoids. The gametophytes (and the underground stems of the sporophytes) of the two genera contain a mycorrhizal fungus within their cells, on which they depend for their nutrition. For further details of the structure and development of *Psilotum* and *Tmesipteris* see Bierhorst's (1971) well-illustrated descriptions.

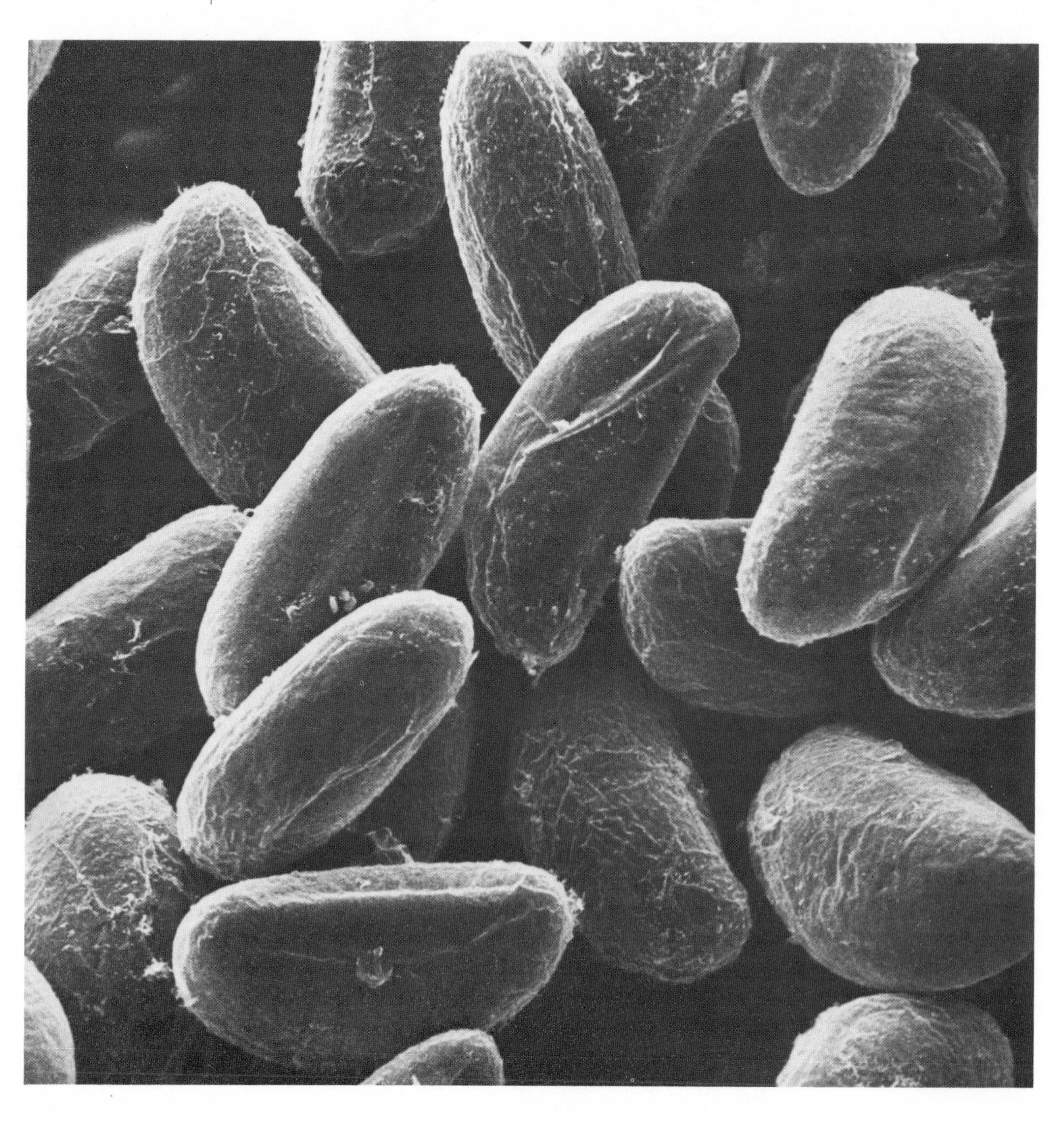

Homosporous and heterosporous plants. They have small one-veined leaves known as microphylls.

Plate 58 X 50

Bisexual cone of *Selaginella kraussiana* A. Br.

There are approximately 700 species of *Selaginella* restricted mainly to the tropics and subtropics although some occur in subarctic regions. *S. kraussiana* is a small creeping plant; some species are climbers and others are upright plants several feet tall.

Selaginella differs from ferns, psilophytes, horsetails and *Lycopodium* in being heterosporous. It shares this feature with gymnosperms, angiosperms and a few other extant pteridophytes—*Isoetes* of the Lycophyta and the water ferns (Marsileales and Salviniales). Heterospory involves the production of two different types of spores—microspores which become the male gametophytes and larger megaspores (macrospores) which grow into female gametophytes. The sporophylls (sporangia-bearing leaves) are spirally arranged to form a cone. Each sporophyll bears a single-stalked sporangium on its upper surface. A microsporophyll bearing a

Plate 59 X 50

Plate 60 X 160

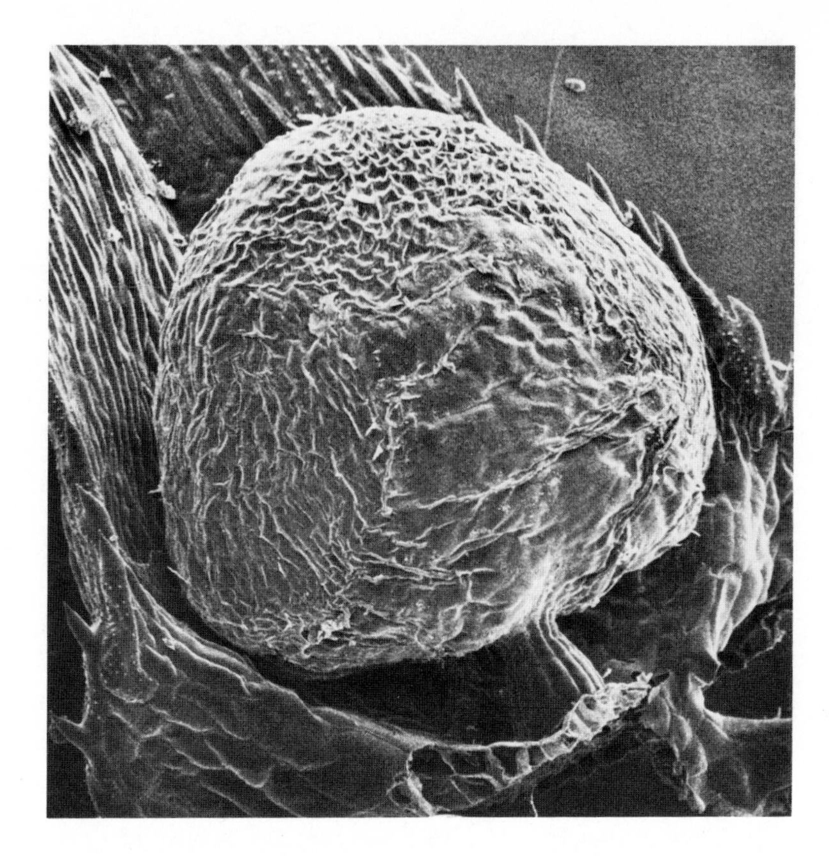

microsporangium is shown in Plate 60. In *S. kraussiana* there is a single megasporangium at the base of the cone (Plate 59). It is larger than the microsporangia (Plate 58) but contains only four megaspores. There are from 1 to 12 megaspores per megasporangium in other species. Each microsporangium contains many microspores. Megaspores and microspores are formed by the meiotic division of the appropriate mother cells.

Microspores develop into multicellular male gametophytes before they are shed from the microsporangium. Plate 61 shows a microspore at the stage at which it is liberated. Similarly, each megaspore undergoes development before it is liberated and archegonia are formed on the upper surface of this female gametophyte. In some cases, fertilisation occurs before female gametophytes are shed. The megasporangium wall partly opens to allow entry of microgametophytes. The spiny wall of the microgametophyte (the old microspore wall) breaks down and motile biciliate sperms are released.

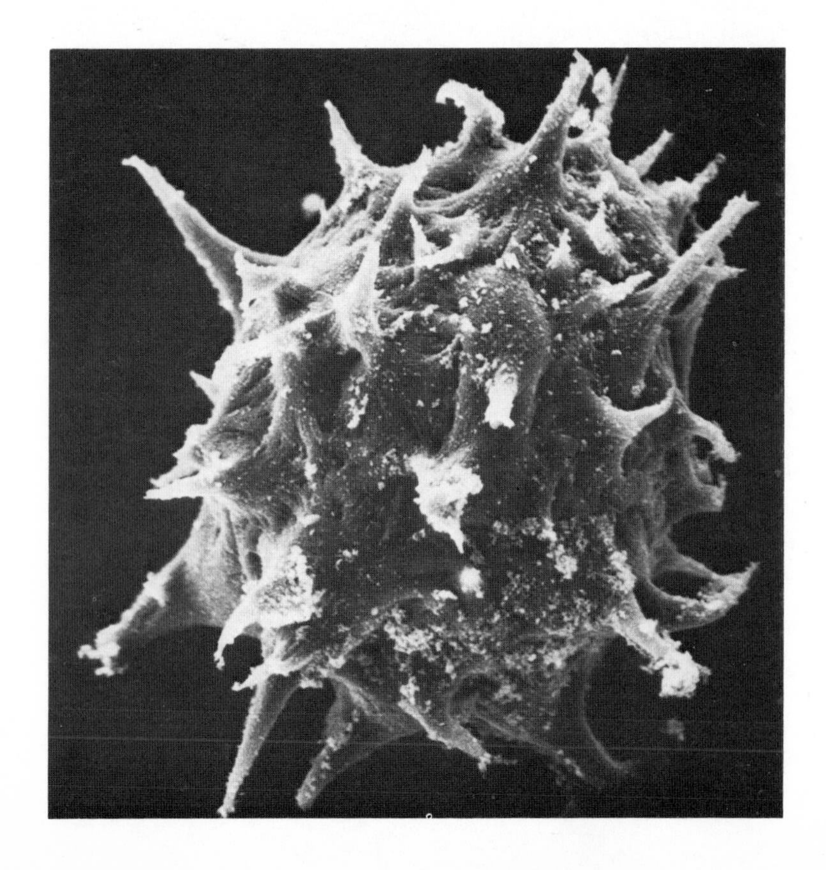

Plate 61 X 1,800

SUBPHYLUM SPHENOPHYTINA

The horsetails. Homosporous plants with jointed stems which contain silica. Leaves are scale-like and sporangia are borne in a cone at the stem apex. Comprises a single extant genus, *Equisetum*.

Stem surface of common horsetail (*Equisetum arvense* L.)

Plate 62

X 1,400

Equisetum is the surviving relic of an ancient and once diverse group of pteridophytes. Whorls of reduced scale leaves are situated at joints on the vertically ribbed stem of this herbaceous plant. Photosynthesis is carried out principally by the green stems.

The small knobs which cover the epidermis of stem and leaves consist of silica ($SiO_2.nH_2O$). Silicon is an essential element for healthy growth of *Equisetum*. It has been suggested that the silica gives strength to the cell walls and that the deposits represent excess silica, which has been absorbed via the roots, but is not required by the plant. The siliceous knobs give a hard texture to the shoots and in colonial times these were used for cleaning and polishing pots and pans. The silica deposits probably account for the high resistance of the plant to fungal attack. The intricate and regular arrangement of the knobs has led Kaufman, Bigelow, Schmid and Ghosheh (1971) to postulate that the silicon may be transported (as silicic acid) through small pores in the outer cell wall, to be deposited beneath the cuticle, but essentially outside the outer epidermal cell wall. The silicic acid becomes silica as a result of dehydration.

The stomatal apparatus is an unusual one, for the two guard cells are not visible externally. They are covered by two subsidiary cells which lie directly above. The four cells are derived from the same stoma mother cell (mesogenous type of development). At an early stage the guard cells are visible externally and are surrounded by the two subsidiary cells, but the latter grow over them.

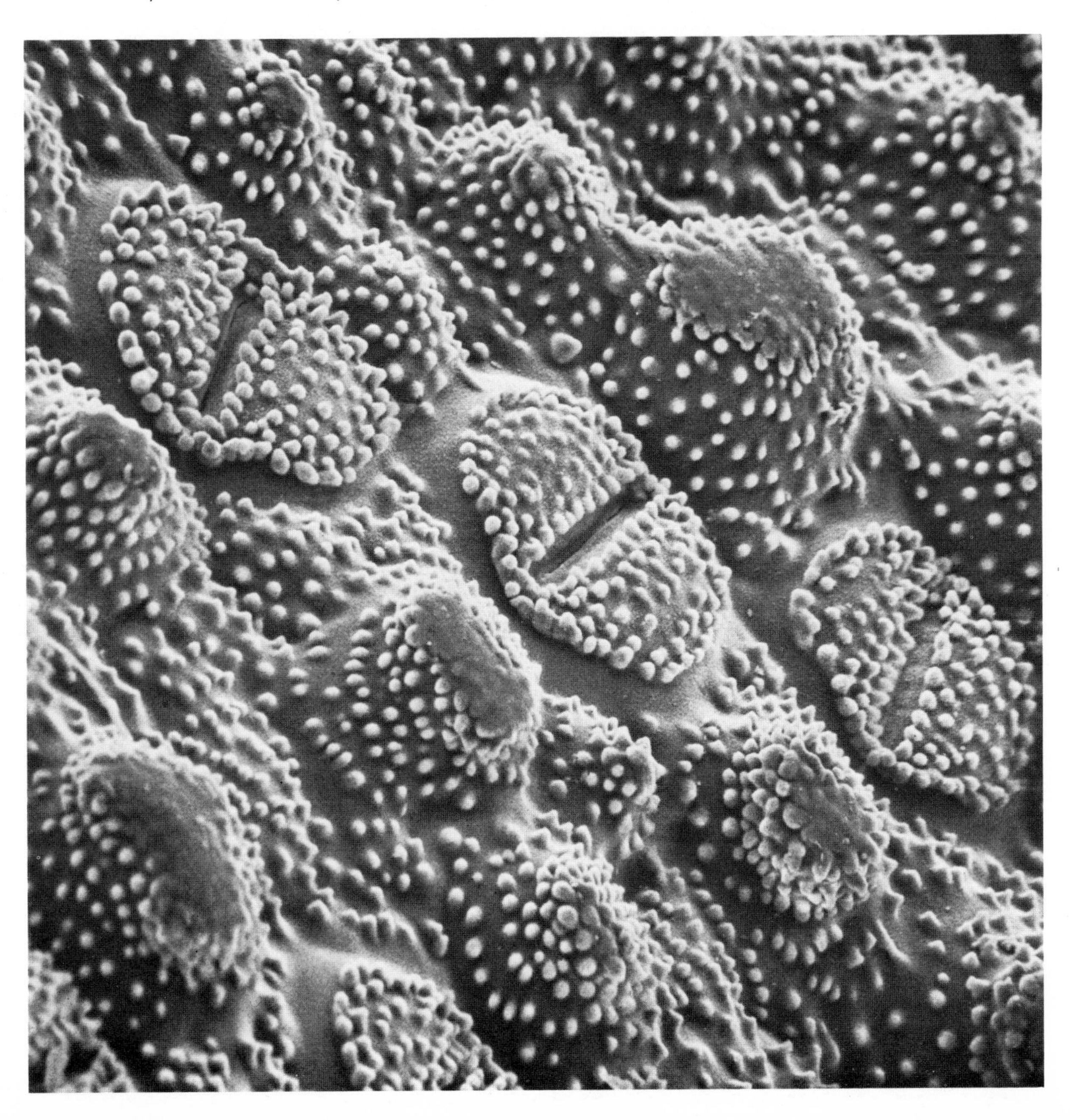

Plate 63 X 70

Sporangia of common horse-tail *(Equisetum arvense)*

The cone of *Equisetum* consists of many whorls of closely packed appendages, termed sporangiophores. Each sporangiophore is rather like an umbrella with a flattened top and a hexagonal outline, which faces the outside of the cone. Extending downwards from the underside of the flattened top are five to ten sporangia which surround the stalk of the sporangiophore. The stalk is attached to the axis of the cone. Each sporangium originates on the outer surface of the sporangiophore but becomes re-aligned during growth. Three sporangia can be seen in lateral view. The outside of the cone is at the top of the photo. The sporangia dehisce along a line parallel to and facing the sporangiophore stalk.

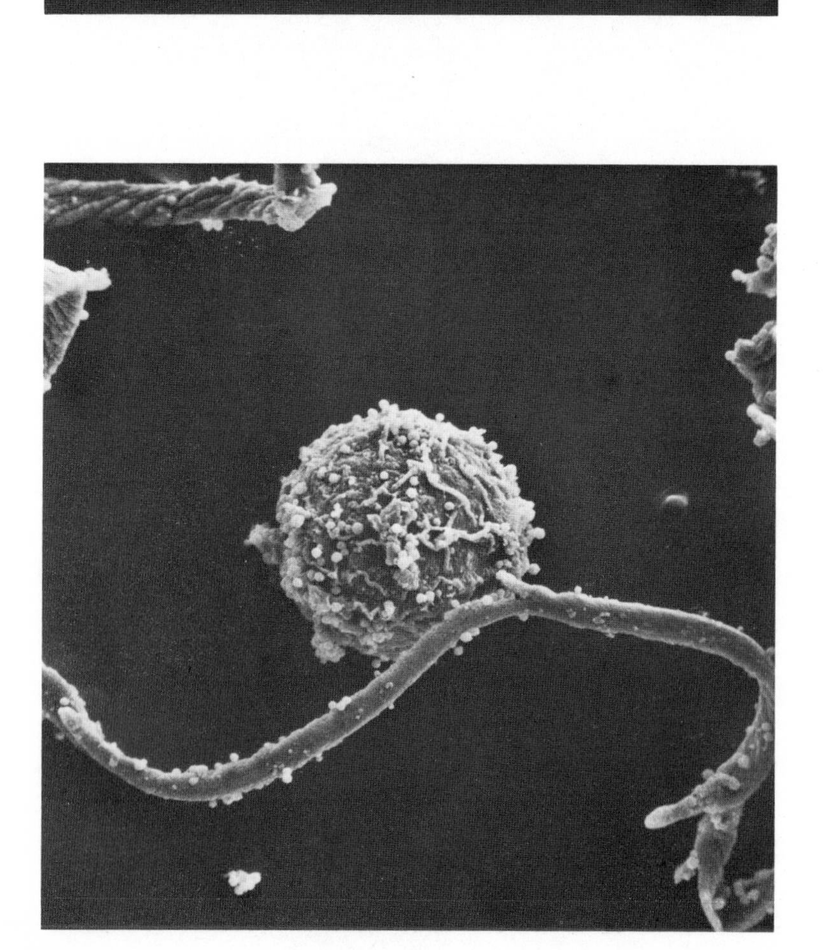

Plate 64 X 600

Spore and elaters of common horsetail (*Equisetum arvense* L.)

The elaters of *Equisetum* are unlike those of *Marchantia* (Plate 36) and other plants because they are formed from the outer layer of the spore wall itself. This layer, the exosporium, is laid down in spiral strips which separate from the rest of the spore wall except at the point of attachment (Plate 65). This more magnified view shows the rope-like structure of the elater, which gives it considerable strength. The elaters have a two-ply structure somewhat analagous to the outer peristome teeth of *Funaria* (Plate 49). In moist air the elaters curl around the spore (Plate 66). At centre right part of the flattened spoon-shaped tip of an elater is visible. In dry weather they stretch out (Plate 64).* The small round spheres which coat the spores and elaters are an inherent part of them and not "foreign" matter. It has been suggested the elaters function primarily to separate spores

Plate 65

X 3,500

from one another, although they often seem to keep them entangled (Bierhorst, 1971). It has also been suggested that in dry weather the spores can be dispersed over a greater distance because the outstretched elaters serve as a parachute.

** This photo shows only one of the two elaters which are attached to each spore. The other became dislodged during pre-treatment for SEM examination.*

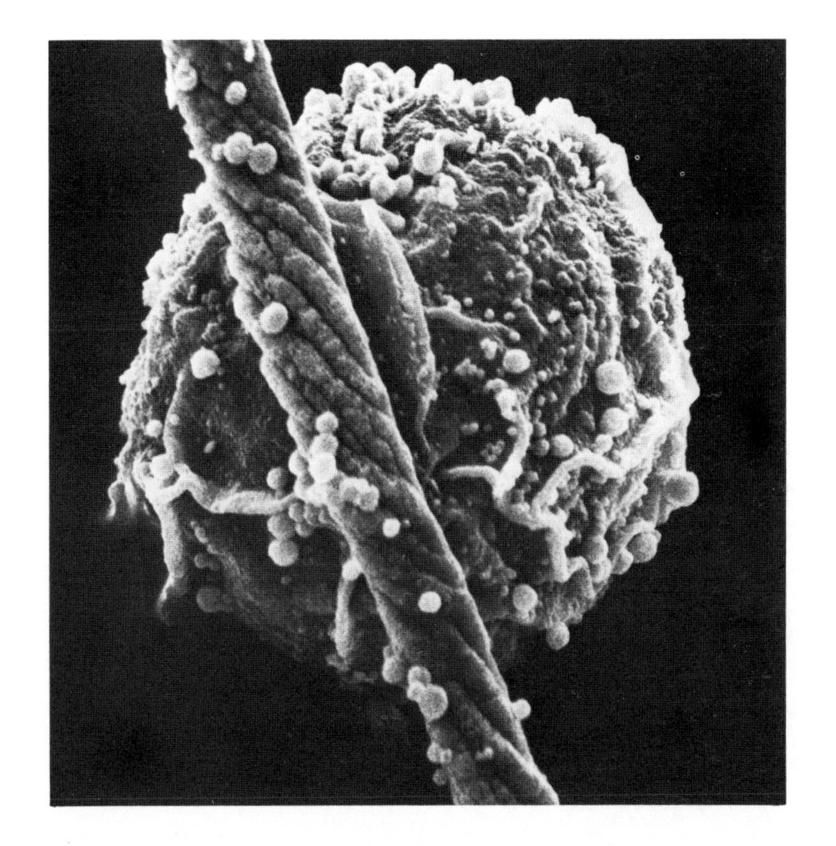

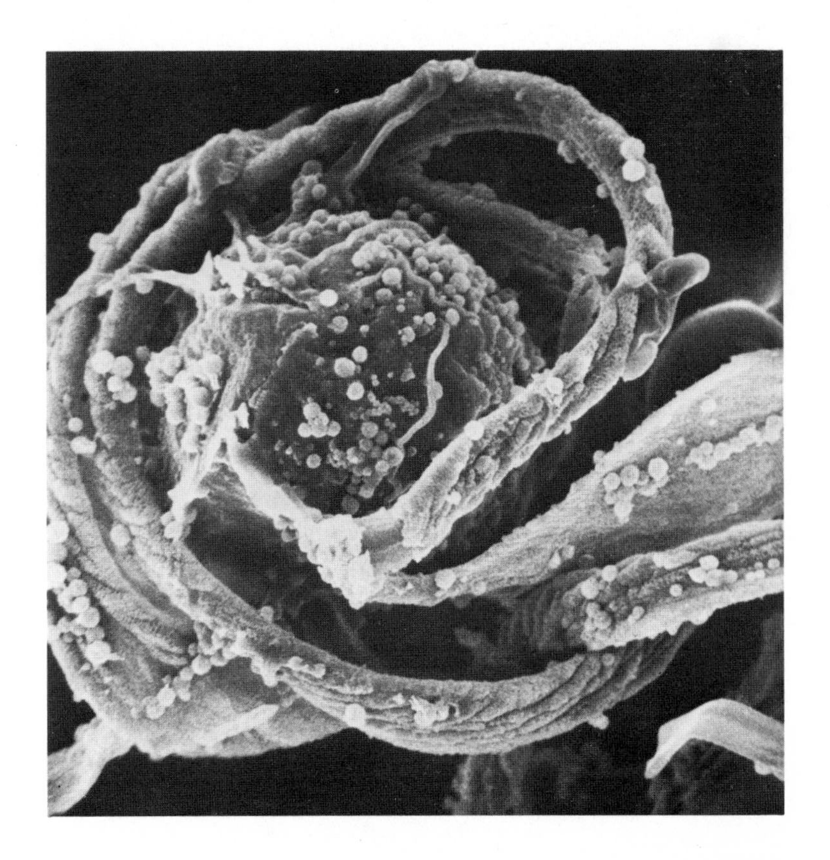

Plate 66

X 2,000

SUBPHYLUM PTEROPHYTINA

Ferns and seed plants. A diverse group with large multiveined leaves known as megaphylls. In some genera the megaphylls have become very reduced.

CLASS FILICINEAE

Ferns. Most are homosporous. The gametophytes are free-living and usually photosynthetic.

Plate 67

X 1,000

Stomata on fern leaflet (*Hypolepis tenuifolia* (Forst. f.) Bernh.)

In this fern, stomata are confined to the lower surface of the leaflets. The walls of the kidney-shaped guard cells that border the stoma are much more heavily thickened than elsewhere. Opening of the stomatal pore occurs when the guard cells enlarge by absorption of water and the thinner, more elastic, walls which do not surround the stoma stretch and bulge outwards and draw with them the thick, comparatively inelastic walls around the stoma. Note the undulating surface and sinuous anticlinal walls (those walls at right angles to the surface) of the epidermal cells. The epidermal cells of fern leaflets contain chloroplasts, in contrast to the epidermis of angiosperm leaves where, with some exceptions (e.g. leaves of some submerged aquatics), chloroplasts are lacking. Guard cells of all types of plants contain chloroplasts. To date, there have been few studies on the structure and development of stomata in ferns.

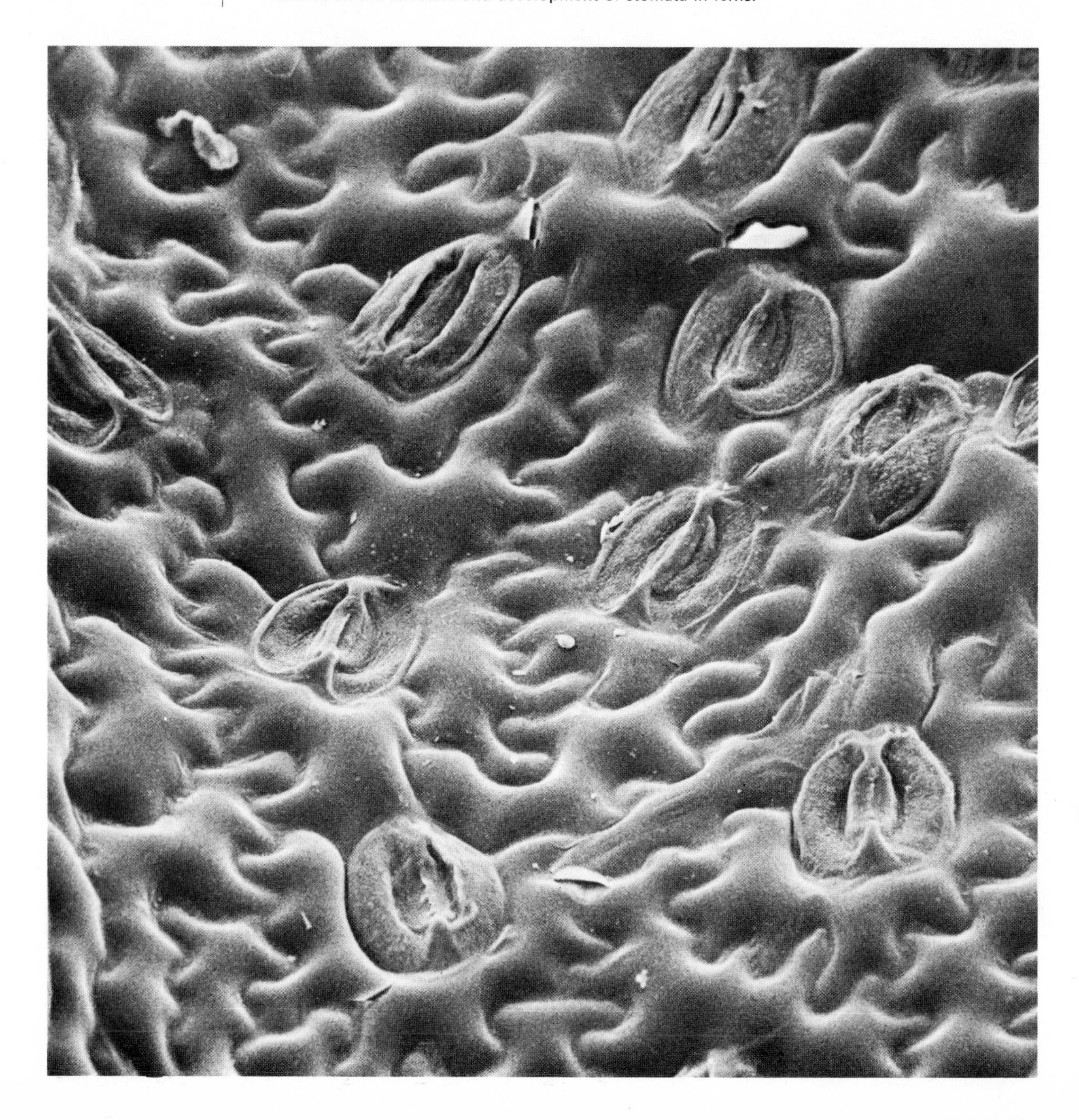

Transverse section of a rhizome of bracken fern (*Pteridium aquilinum* var. *esculentum* (Forst. f.) Kuhn)

Plate 68
X 80

Part of the vascular system is illustrated in this section of an underground stem (rhizome) of bracken fern. Secondary vascular tissue (see Plate 114) is not present in ferns, with the exception of *Botrychium* (Bierhorst, 1971). There are a number of bundles of vascular tissue in the stem. They are interconnected to form a network. Each is termed a meristele and is surrounded by a single layer of cells known as the endodermis (**E**). Within the meristele, primary xylem (water-conducting tissue) is surrounded by primary phloem (food-conducting tissue). The smaller cells of the first-formed primary xylem occur as groups (**X**) surrounded by the later-formed metaxylem. The endodermis encloses a few layers of pericycle parenchyma cells.

V—a vessel of the metaxylem, in which part of a sloping end wall is visible;

P—a sieve cell of the primary phloem

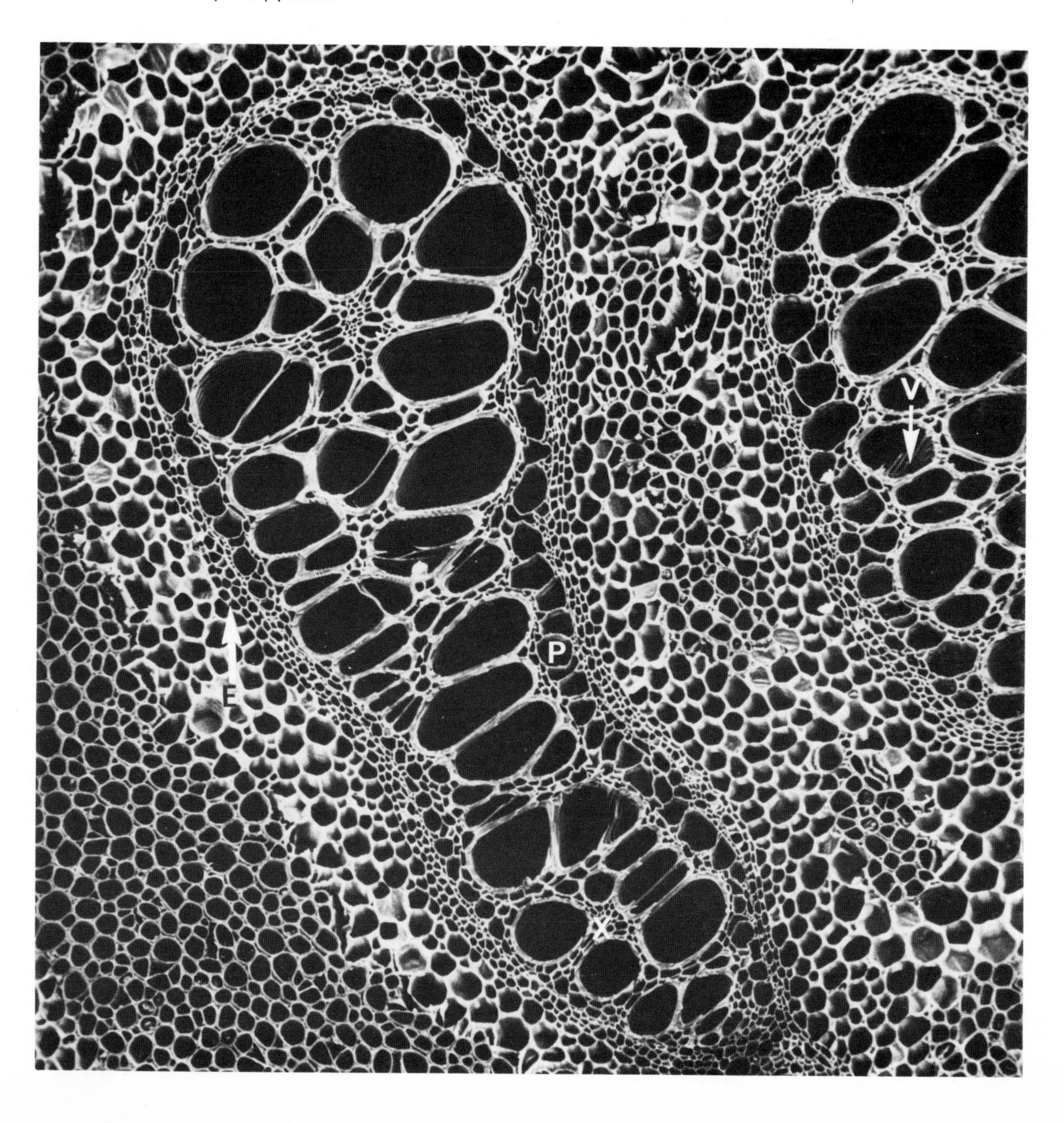

Plate 69

X 3,400

Vessels and sieve cells in a meristele of bracken fern

This is a more magnified view of part of Plate 68. Bracken is unusual among ferns in having vessels in the metaxylem, rather than tracheids (see Plates 103–106). Vessels have perforations (**P**) in their end walls and these do not occur in tracheids, which have only pits. The sloping end wall, which is only partly shown in the photo, has the perforations aligned one above the other, to form a scalariform (ladder-like) perforation plate.

The sieve cells in the primary phloem contain numerous sieve areas (**S**) on their lateral walls, which are long and oblique. There are parenchyma cells (**PA**) between the vessels and sieve cells.

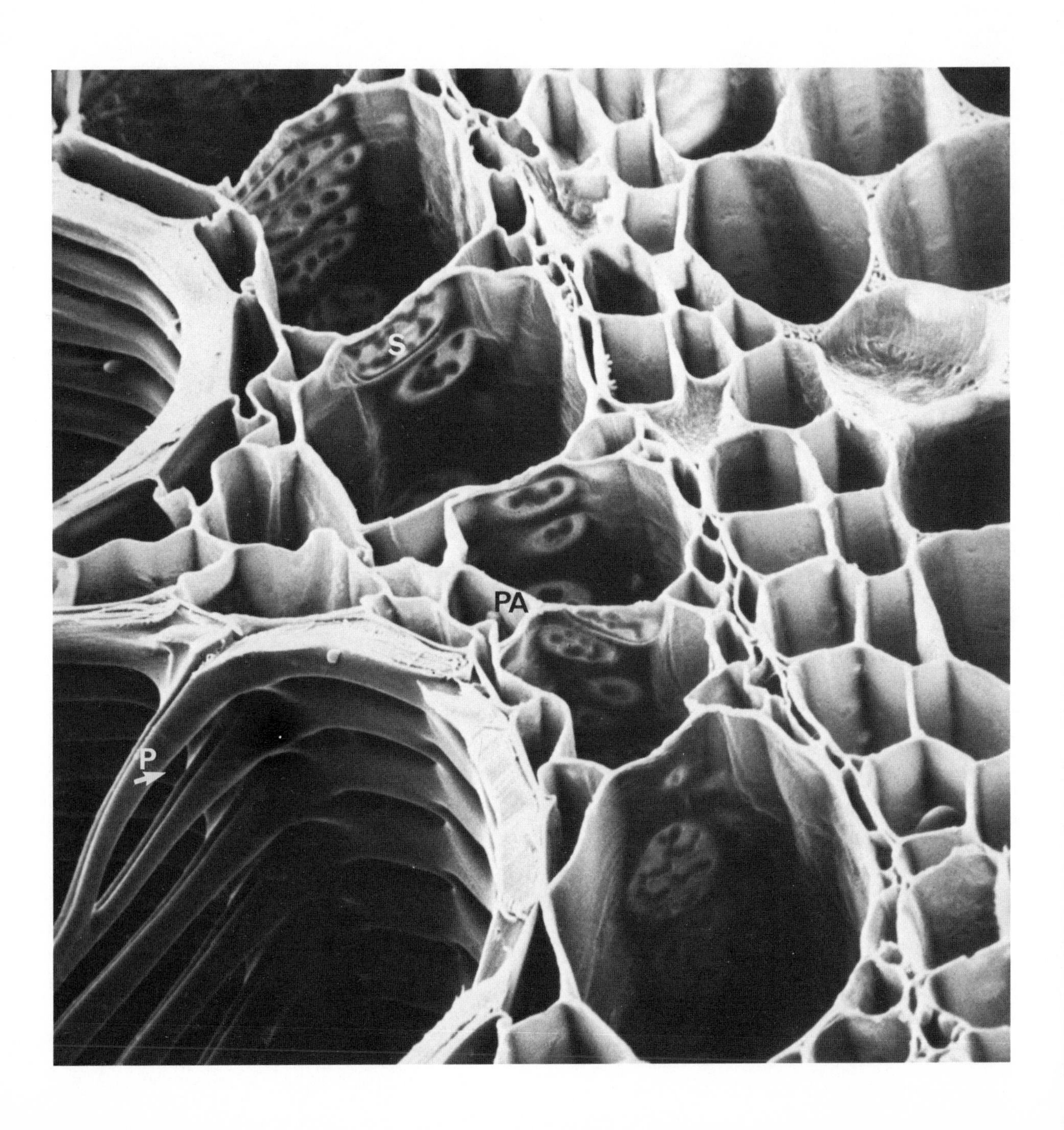

Sieve areas in sieve cells of bracken fern

Plate 70

X 6,700

The spherical protuberances on the sieve areas on the walls of the sieve cells have been termed "refractive spherules" (Esau, 1969). It has been reported that they are lodged against, and partly sunken into, the sieve pores. The sieve pores, through which food materials pass from one cell to another, are thus concealed by the spherules. The sieve pores occur in clusters known as sieve areas. Under the electron microscope, the refractive spherules appear as organelles with a double membrane and an electron dense centre (Esau, 1969). They also occur dispersed in the cytoplasm of the sieve cells, which at maturity are without nuclei.

The spherules do not occur in the phloem of gymnosperms and angiosperms. Chemical tests indicate they are proteinaceous.

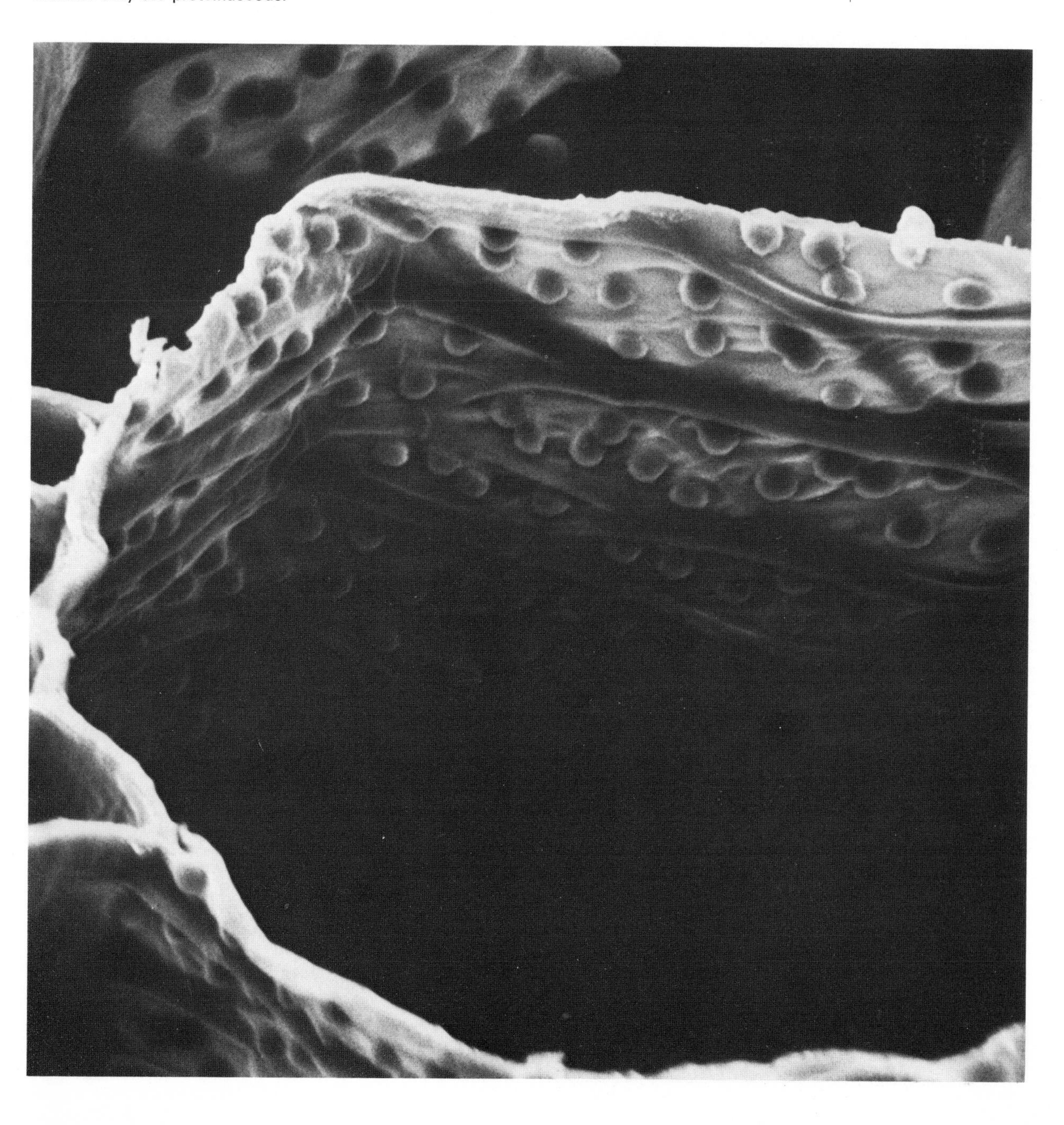

Plate 71
X 2,000

Tracheids in the leaf stalk of a fern *(Hypolepis tenuifolia)*

In this view of part of a transverse section, the large cells are xylem tracheids. Tracheids are dead at maturity and function as water-conducting tissue. The elongated pits in the side walls of the tracheids are arranged one above the other, to form ladder-like series called scalariform pitting (from the Latin *scalaria*—a staircase).

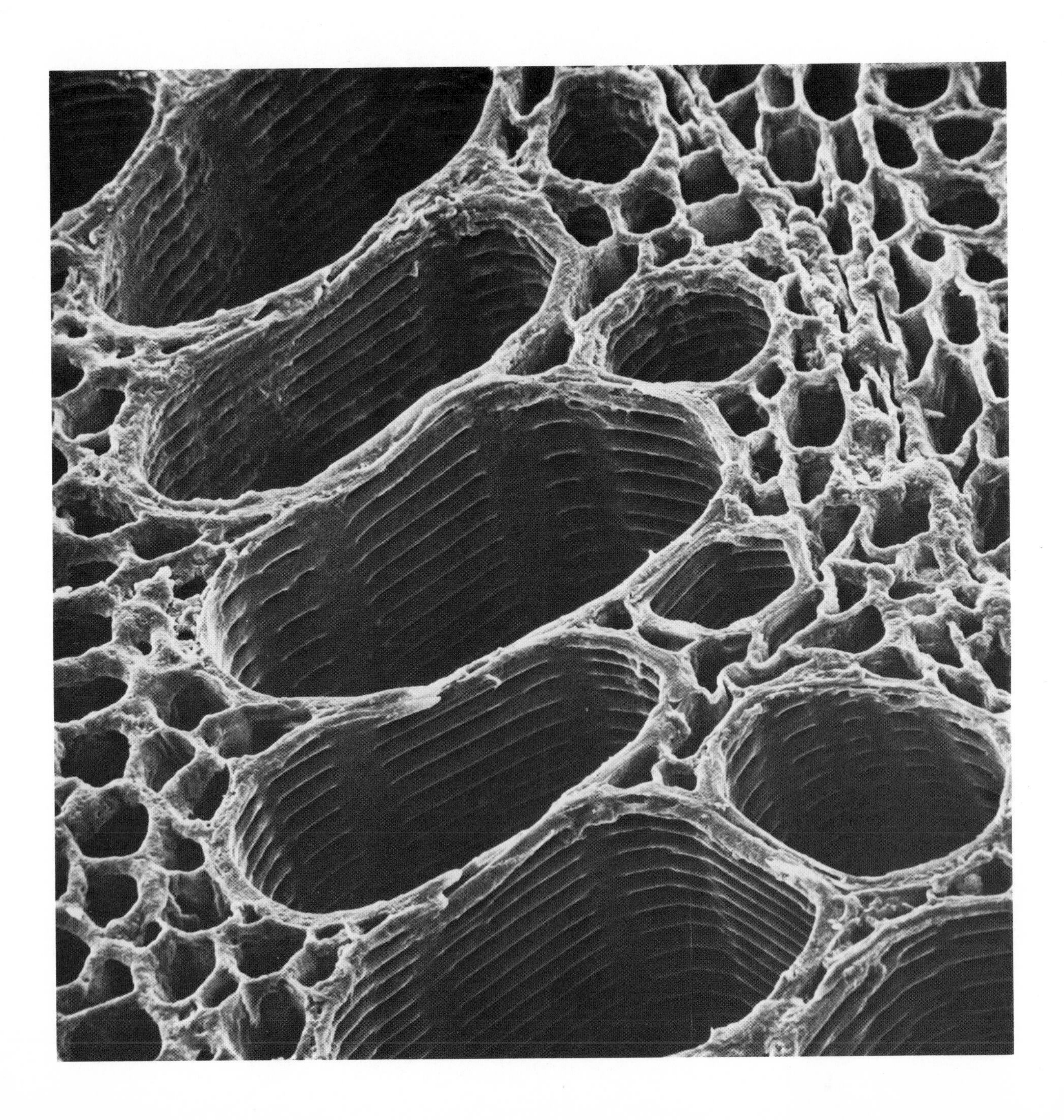

Plate 72
X 5,000

Tracheid wall structure in a fern (*Phymatodes diversifolium* (Willd.) Pic. Ser.)

This illustrates part of two adjacent tracheids in the rhizome of *Phymatodes*. The scalariform pits (see Plate 71) on the side walls are shown in face and sectional view. It has been noted (Plate 69) that whereas vessels have perforations in their end walls, tracheids have only pits. Pits represent thinner regions of the wall. These regions are thin enough to enable water to pass readily from one tracheid to another.

The pit membrane (**M**) can be clearly seen in sectional view.

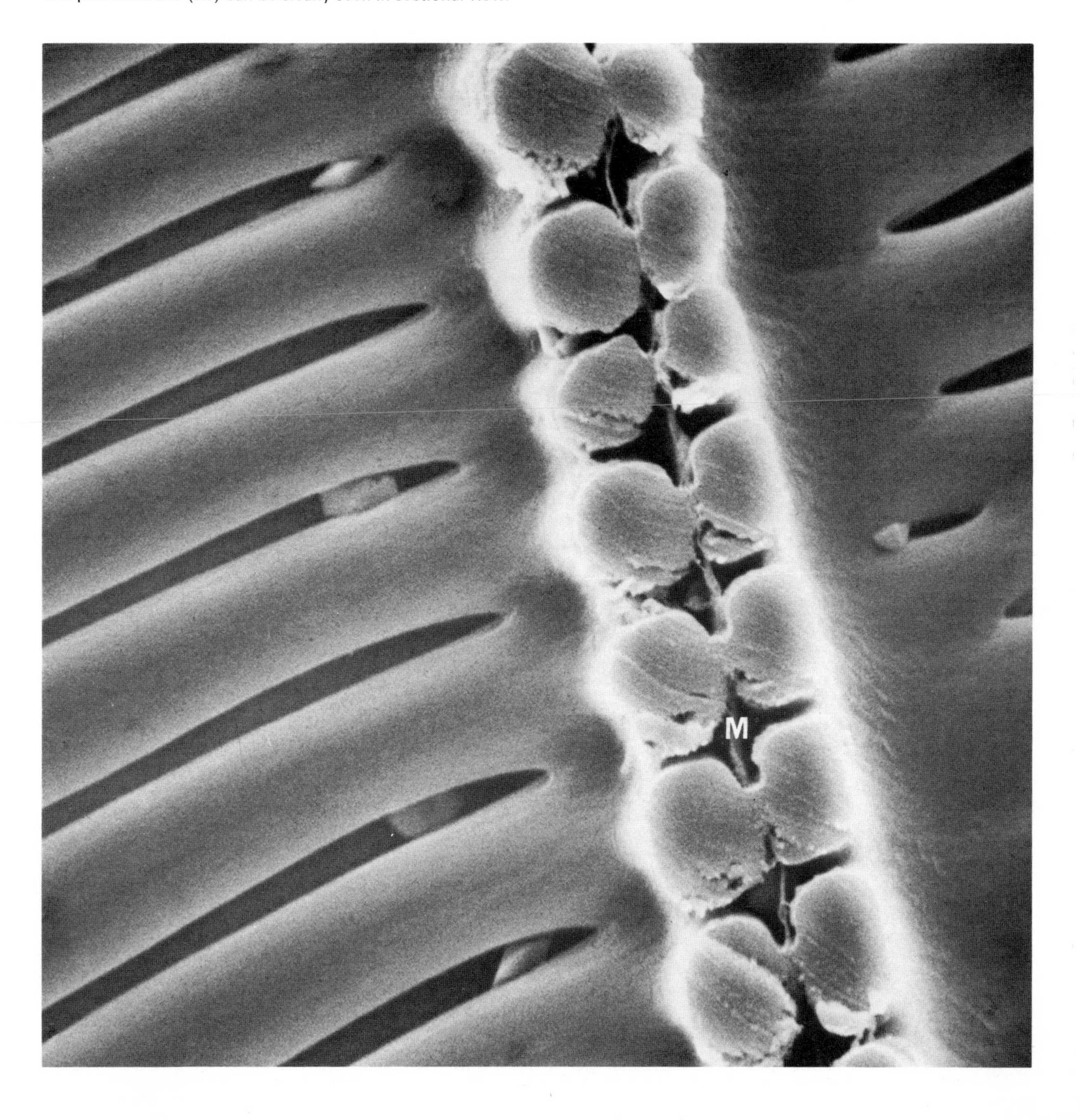

Plate 73
X 220

Sorus on leaflet of New Zealand silver tree fern (*Cyathea dealbata* (Forst. f.) Swartz)

The sporangia are arranged in hemispherical groups (sori) in two rows on the silver-coloured underside of the leaflets. These immature sporangia are still packed tightly together.

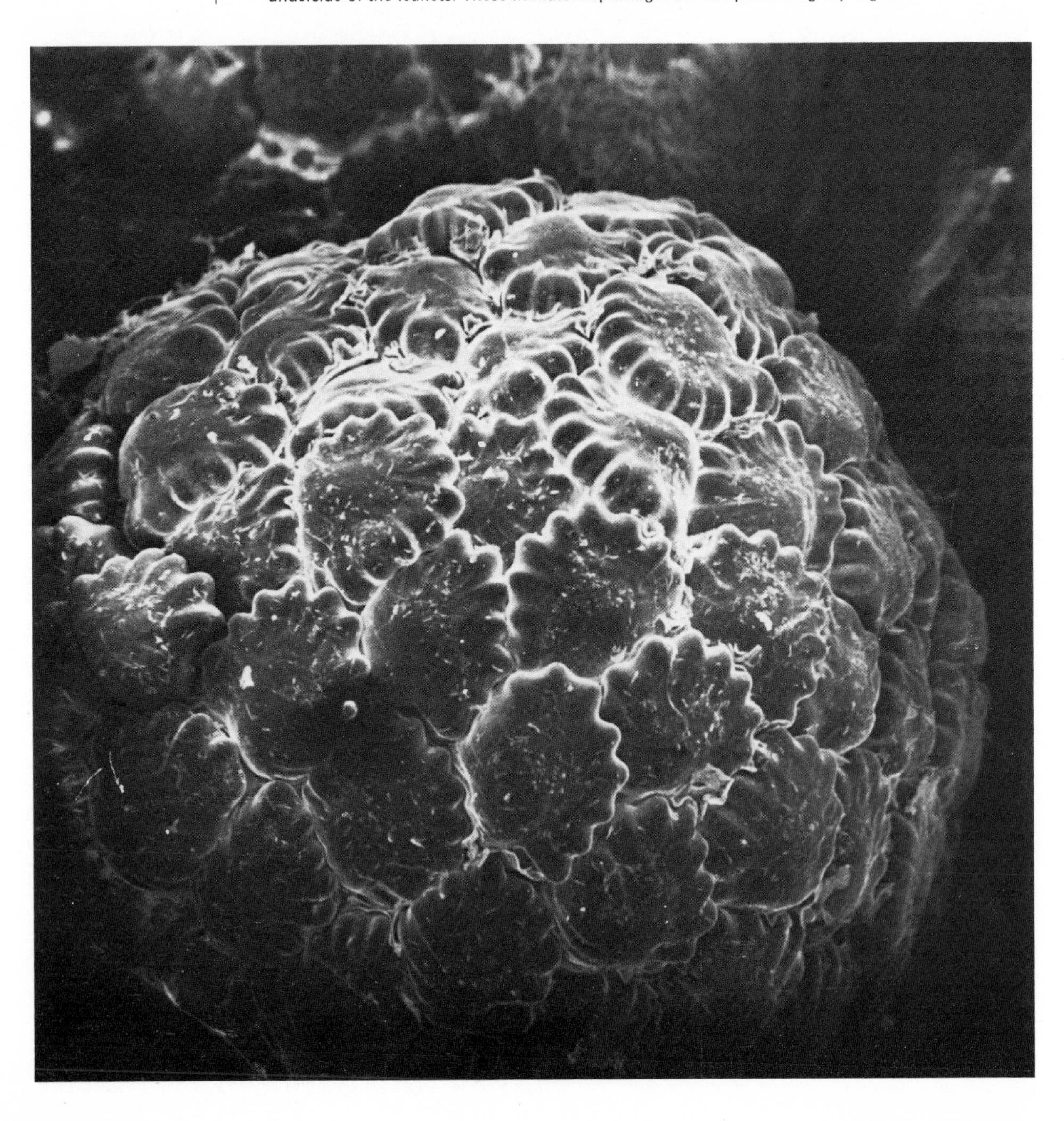

Sporangia on pinna of fern *(Hypolepis tenuifolia)*

The sporangia are arranged in clusters on either side of the midrib, on the underside of the leaflets. In this particular fern, not all sporangia within an individual group (sorus) are at the same stage of development.

Plate 74
X 40

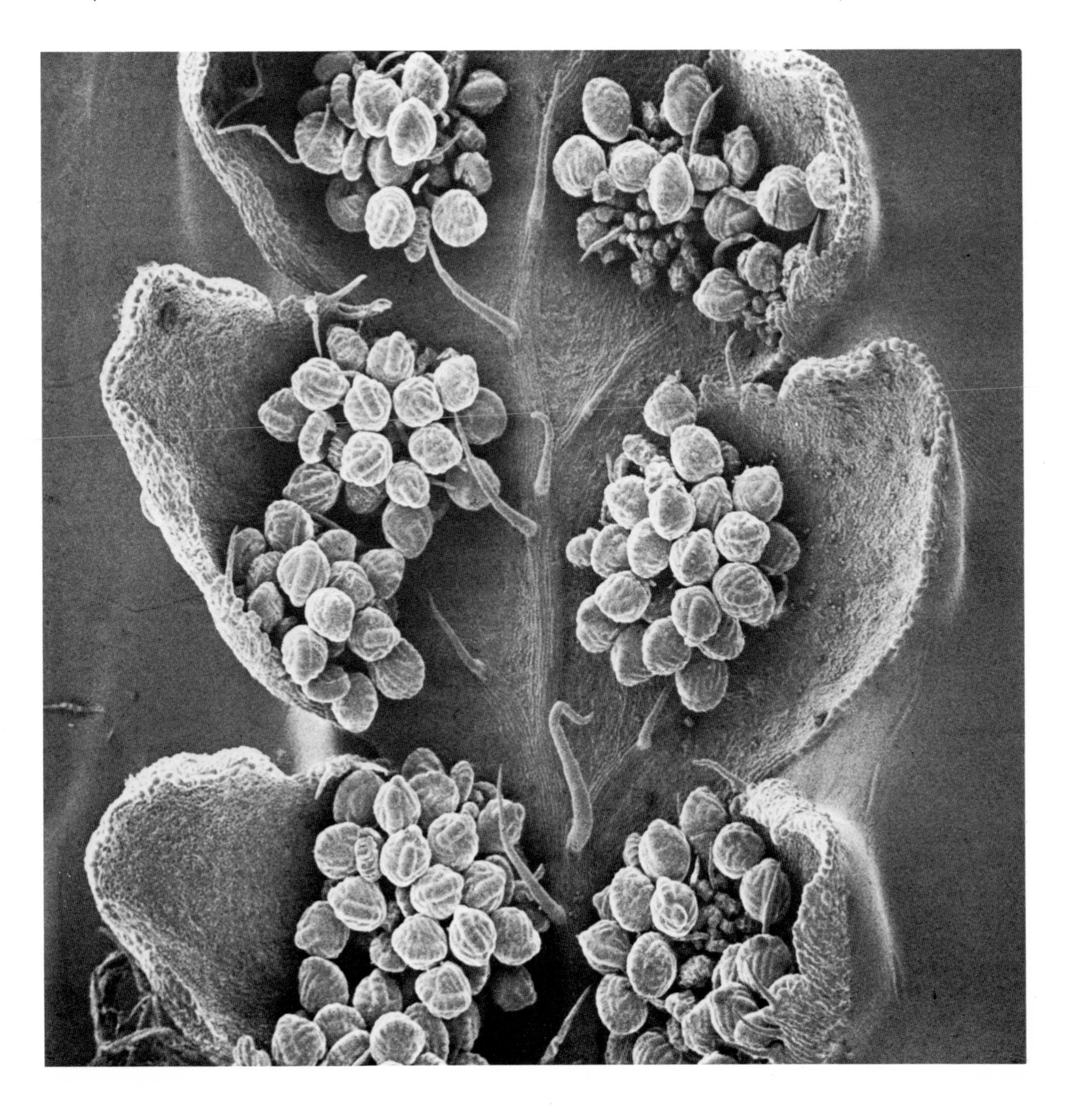

Plate 75

X 280

Fern sporangia *(Hypolepis tenuifolia)*

The sporangium wall is one cell thick. An arc of more prominent wall cells forms the annulus, which runs from the stalk (not visible) and over the top of the sporangium to about halfway back to the other side of the stalk, where it meets the stomium ((**S**), Plate 75), a zone of weaker cells. The opening of the sporangium and liberation of the spores requires dry conditions. Each annulus cell has the innermost wall and the two anticlinal walls (at right angles to the surface) between adjacent annulus cells heavily thickened and is filled with water. As water evaporates from the annulus cells there is a decrease in volume and the thin part of the wall becomes sucked inwards. This sets up strains, the sporangium splits in the stomium region and the split continues sideways across the unspecialised wall cells. This produces an inverted cup-shaped upper part to the sporangium, containing most of the spores (Plate 76). It is held to the rest of the sporangium below it by the annulus cells on the side where they extend to the stalk. As the annulus cells lose

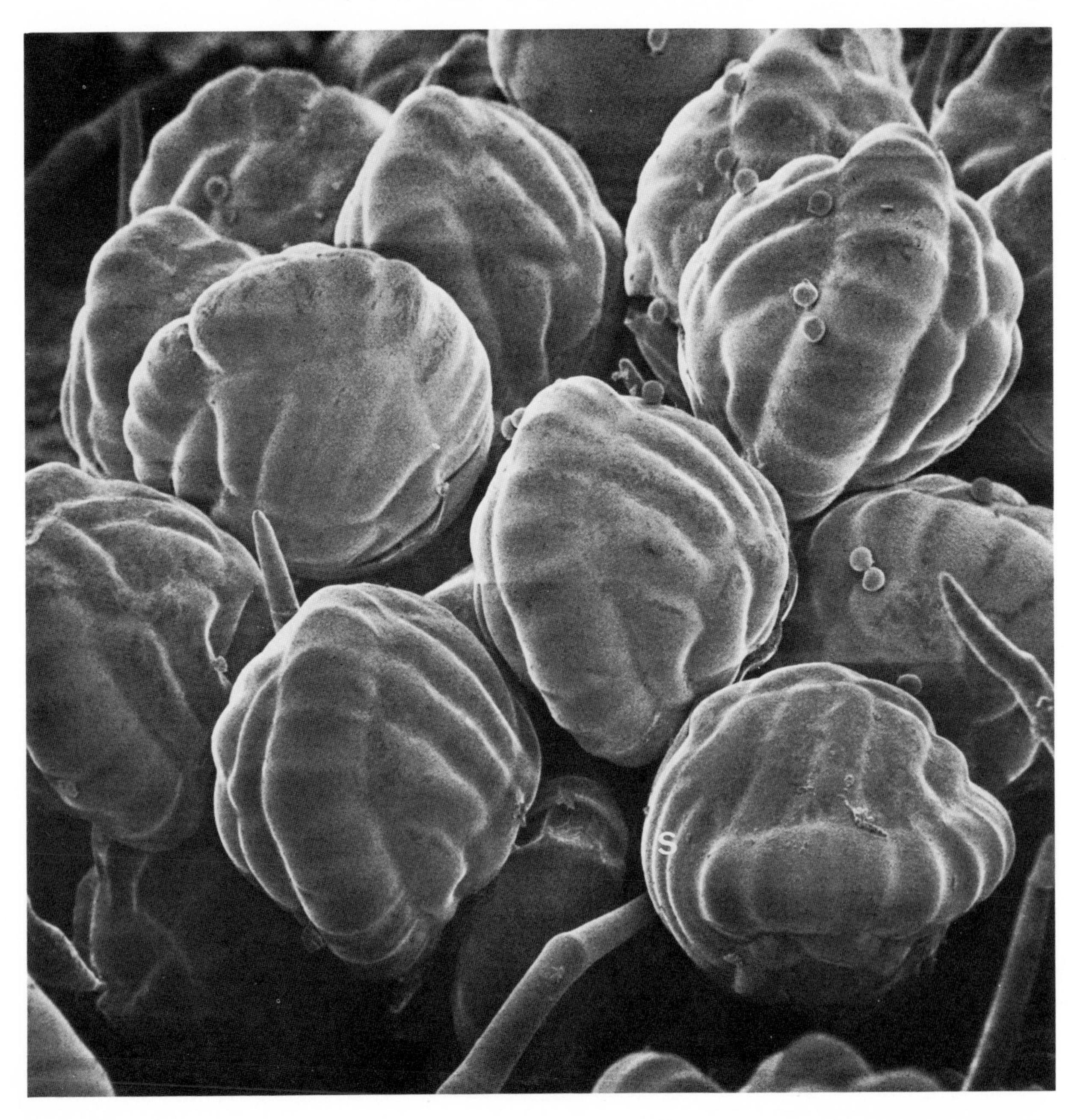

more water, their thin outer walls shrink further inwards and the two thickened anticlinal walls of each cell come closer together. The whole annulus therefore bends slowly backwards and carries with it the cup containing the bulk of the spores. This movement places the water in the annulus cells under tension, due to the elasticity of the thick walls and their tendency to return to their original shape. The tension reaches a level such that the cohesive force of the water molecules is exceeded and with explosive speed liquid becomes gas and water vapour is formed. This releases the tension and the cell returns to its former shape. Movement in one cell sets off others and the annulus with its attached cup of spores is catapulted forwards with considerable speed and spores are flung out. If the dehisced sporangium is then moistened, so that the annulus cells again fill with water, the dehiscence mechanism will be repeated as drying occurs.

Plate 76
X 380

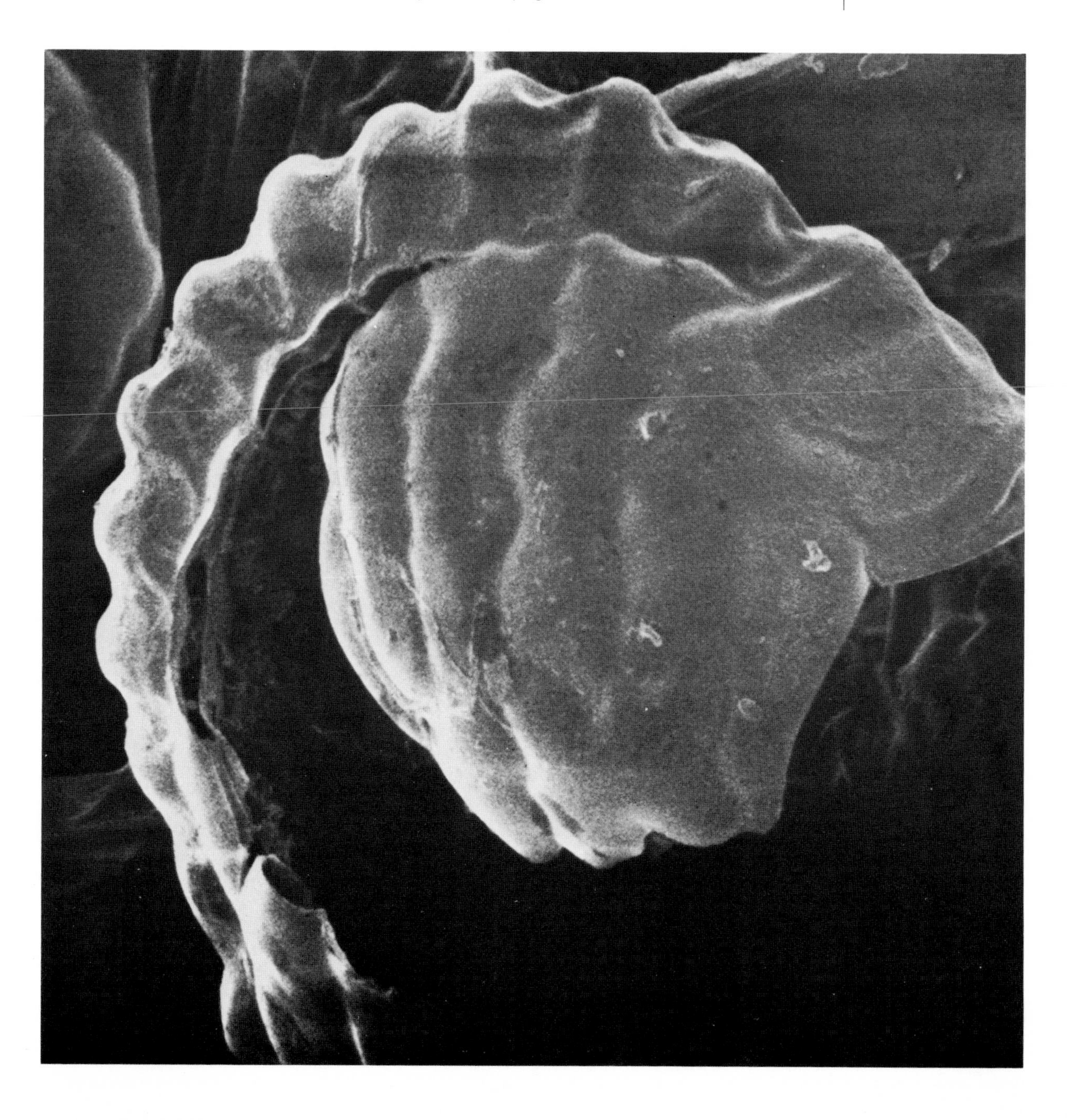

Plate 77 X 1,700

Fern spore (*Blechnum discolor* (Forst. f.) Keys)

Fern spores resemble pollen grains (Plates 91, 92, 143–156) in having distinctive sculpturing on their walls, which frequently allow identification to a generic, or even specific, level. For example, the New Zealand species of *Blechnum* can be identified solely on the basis of differences in spore structure (Harris, 1955).

The spherical or oval spores of *Blechnum discolor* have a thin exine and this is covered by a layer known as the perispore. The perispore is a thin clear envelope which forms a continuous cover around the spore and it is thrown into crest-like ridges (Harris, 1955) as shown.

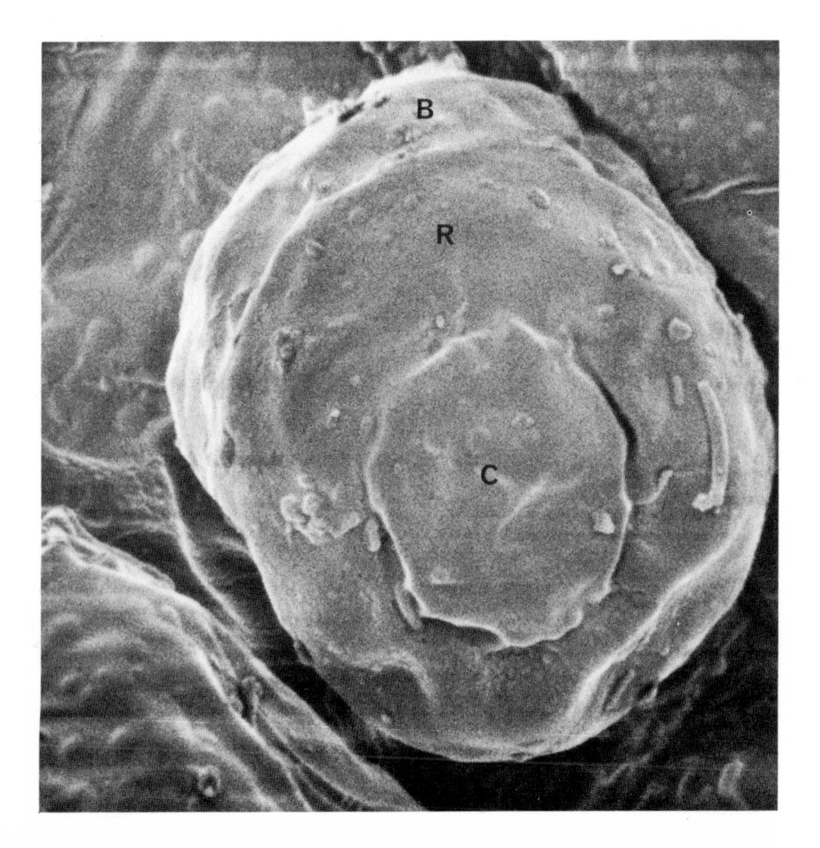

Plate 78 X 2,600

Antheridium on fern prothallus (*Blechnum capense* (L.) Schlecht.)

Antheridia are found on the underside of the small thallose heart-shaped prothallus, which grows to approximately 1 cm in width. Antheridia appear before the prothallus reaches full size. The most advanced fern groups, of which *Blechnum* is an example, have antheridia in which the male gametes are enclosed by only three wall cells. There is a round, flat or somewhat concave basal cell (**B**) on top of which is a ring-shaped cell (**R**) resembling a tube in the older type of automobile tyre. The top of this is plugged by a smaller cap cell (**C**). The development of the antheridium of *Blechnum* is described and illustrated by Bierhorst (1971).

Plate 79 X 1,800

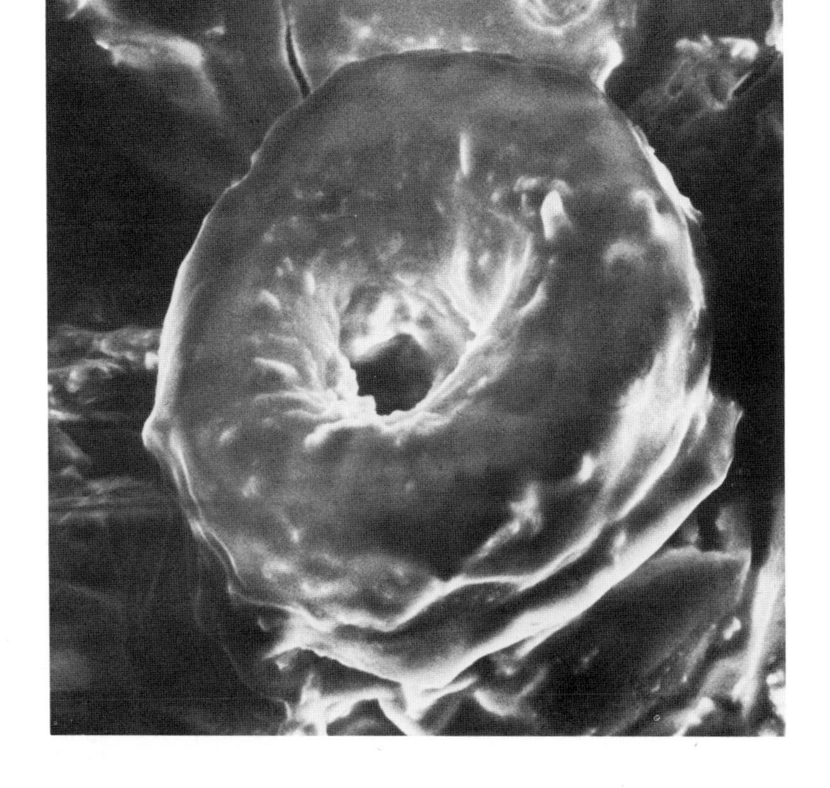

Empty antheridium on fern prothallus *(Blechnum capense)*

Wet conditions are needed for dehiscence. The three wall cells (Plate 78) enclose up to 32 sperm mother cells. Within each mother cell is a developing sperm. The walls of the mother cells become mucilaginous, absorb water and the contents of the antheridium swell. Observations with the light microscope and SEM confirmed that dehiscence resembles that described for a different species of *Blechnum* and other Blechnaceae by Stone (1961, 1969). The antheridial contents rupture the inner wall of the cap cell, enter it and the sperms (either motile, or as yet non-motile) emerge from the antheridium one at a time through a "pore" in the outer wall of the cap cell. The "pore" represents a tear in the outer wall of the cap cell, which extends about a quarter of the way across the top to one side of the centre of the outer wall. In this photo the remains of the cap cell have disappeared.

In many ferns dehiscence is of a different type in which the cap cell unhinges at one side and folds outwards; in others the antheridial contents push the cap cell off the antheridium.

Plate 80 X 8,200

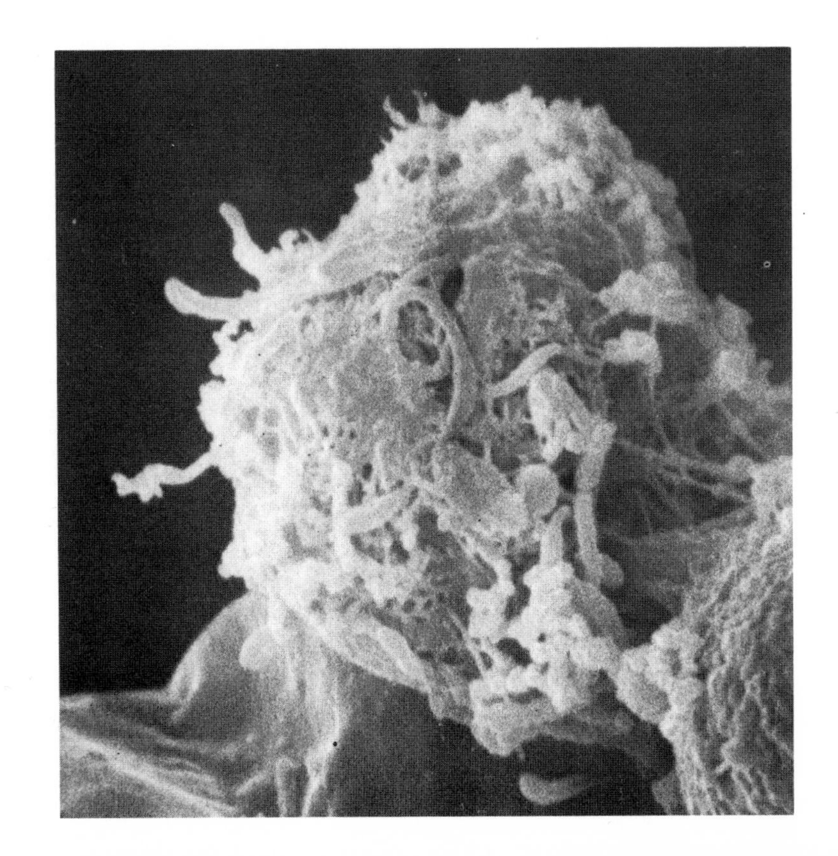

Sperm mass outside antheridium of fern prothallus *(Blechnum capense)*

The sperms leave the antheridium one by one and, if they are not motile at this stage, collect in a mass along with the remains of the mother cell walls, outside the antheridium, until they become motile. Each sperm is a spirally coiled structure consisting mostly of nuclear material and has a large number of cilia (not visible in photo) in contrast to the two cilia in bryophytes (liverworts, mosses and hornworts).

Plate 81
X 800

Archegonia on fern prothallus *(Blechnum capense)*

Archegonia develop later and are closer to the apical notch of the prothallus than the antheridia, which occur among the rhizoids. Only the necks of the archegonia protrude from the lower surface of the prothallus. At maturity the neck is four to six cells in height, four cells in perimeter and encloses a neck canal cell containing two nuclei. In the small venter of the archegonium, which is embedded in the prothallus, there is a ventral canal cell and beneath it the ovum. Before fertilisation occurs, the neck cells at the tip spread apart and the neck canal and ventral canal cells break down, to form a passage-way for the sperm to reach the egg. Note how the necks of older archegonia are curved.

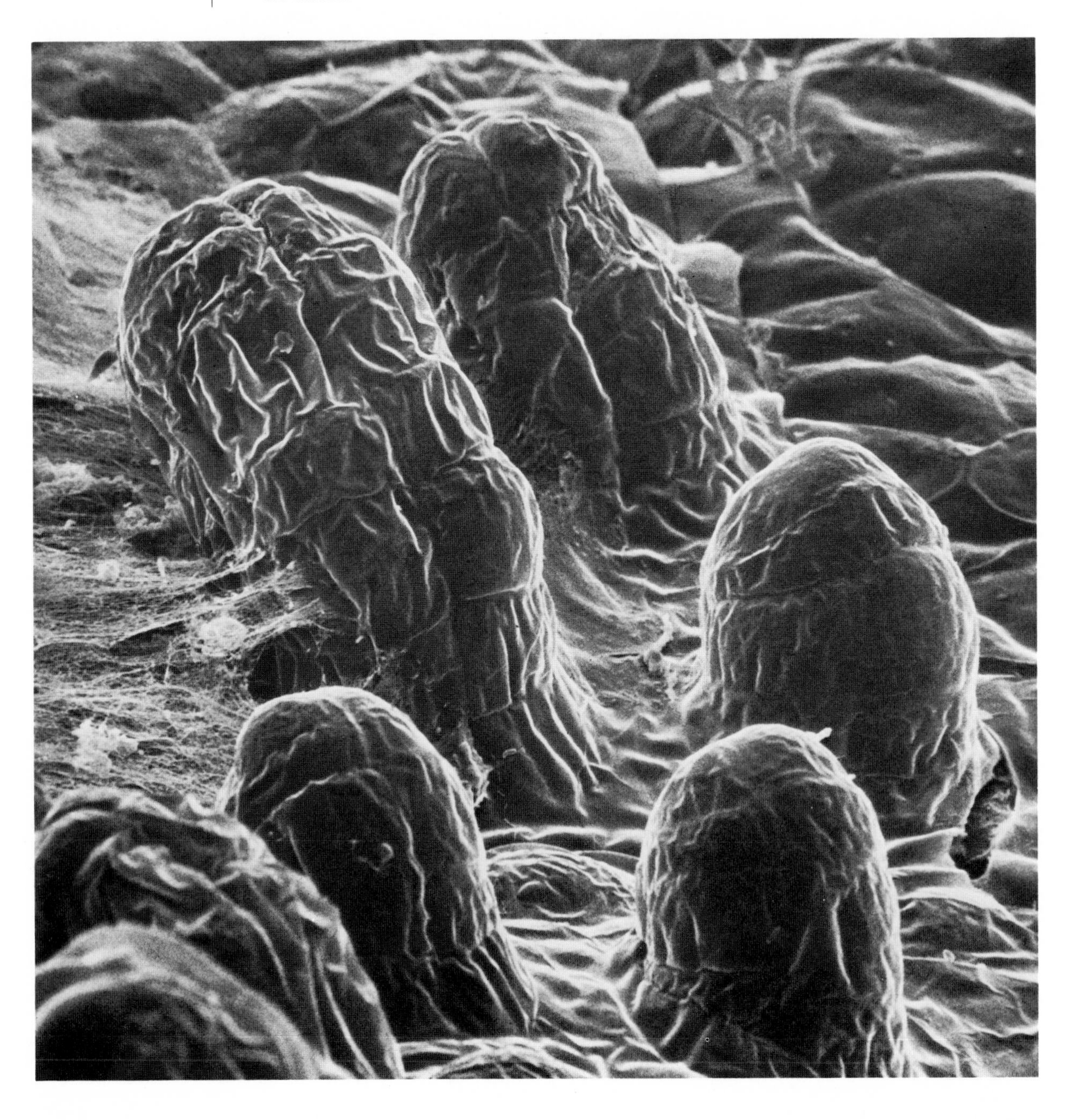

CLASS CONIFERINAE

The conifers. They have abundant secondary vascular tissue. Male gametes do not have flagella. Conifers are the most familiar group of gymnosperms, which are characterised by having ovules which are not enclosed.

Conifer wood in transverse, radial longitudinal and tangential longitudinal section (*Sequoia sempervirens* (D. Don) Endl.)

Plate 82
X 90

At the top of the photo the wood from a stem of redwood is shown in transverse section. Rows of tracheids (**T**) alternate with smaller parenchyma ray cells (**R**). In the tangential longitudinal section at lower right it can be seen that the ray cells (**R**) are in single rows from 2 to 16 cells in height. In the radial longitudinal section (a section which runs from the centre towards the outside of the stem) at lower left, some of the ray cells (**R**$_1$) have been cut open. Redwood trees are the tallest known (*see National Geographic Magazine*, July 1964).

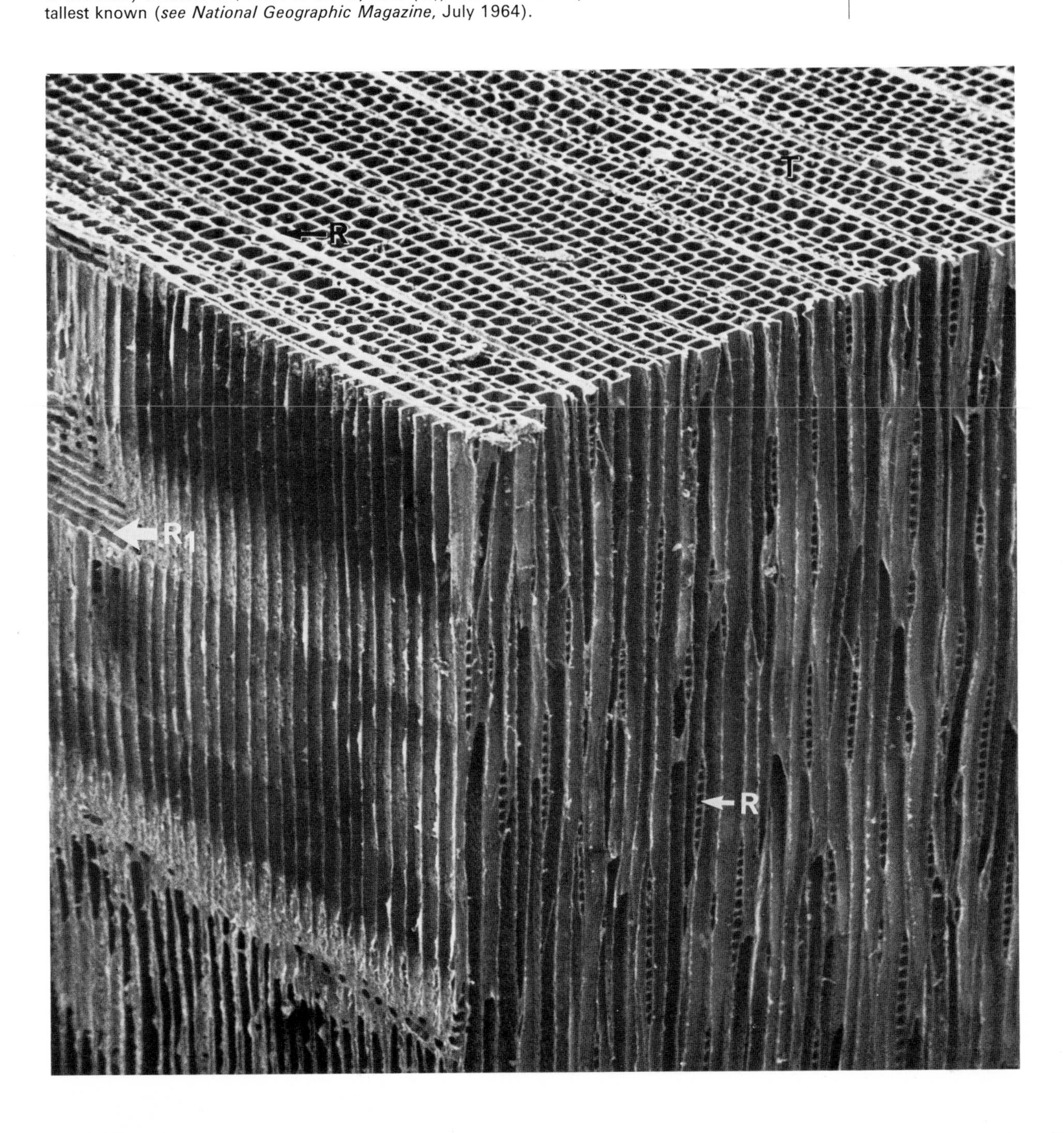

Plate 83
X 1,600

Bordered pits in redwood *(Sequoia sempervirens)*

This is a close-up view of part of the tangential longitudinal section shown in Plate 82. Bordered pits (**B**) are visible in surface view on the radial walls of a tracheid. The lining on the inside wall of this tracheid has spherical protrusions on it and is termed the "warty layer". It has been suggested that the warty layer represents the atrophied remains of the protoplasm. Half-bordered pit-pairs (**H**) can be seen in part sectional view between a tracheid and ray cells. A border is present on the tracheid wall but not on the wall of the ray parenchyma cell. For details of the wood structure of *Pinus* and other plants, as seen with the SEM, the reader is referred to Meylan and Butterfield (1972).

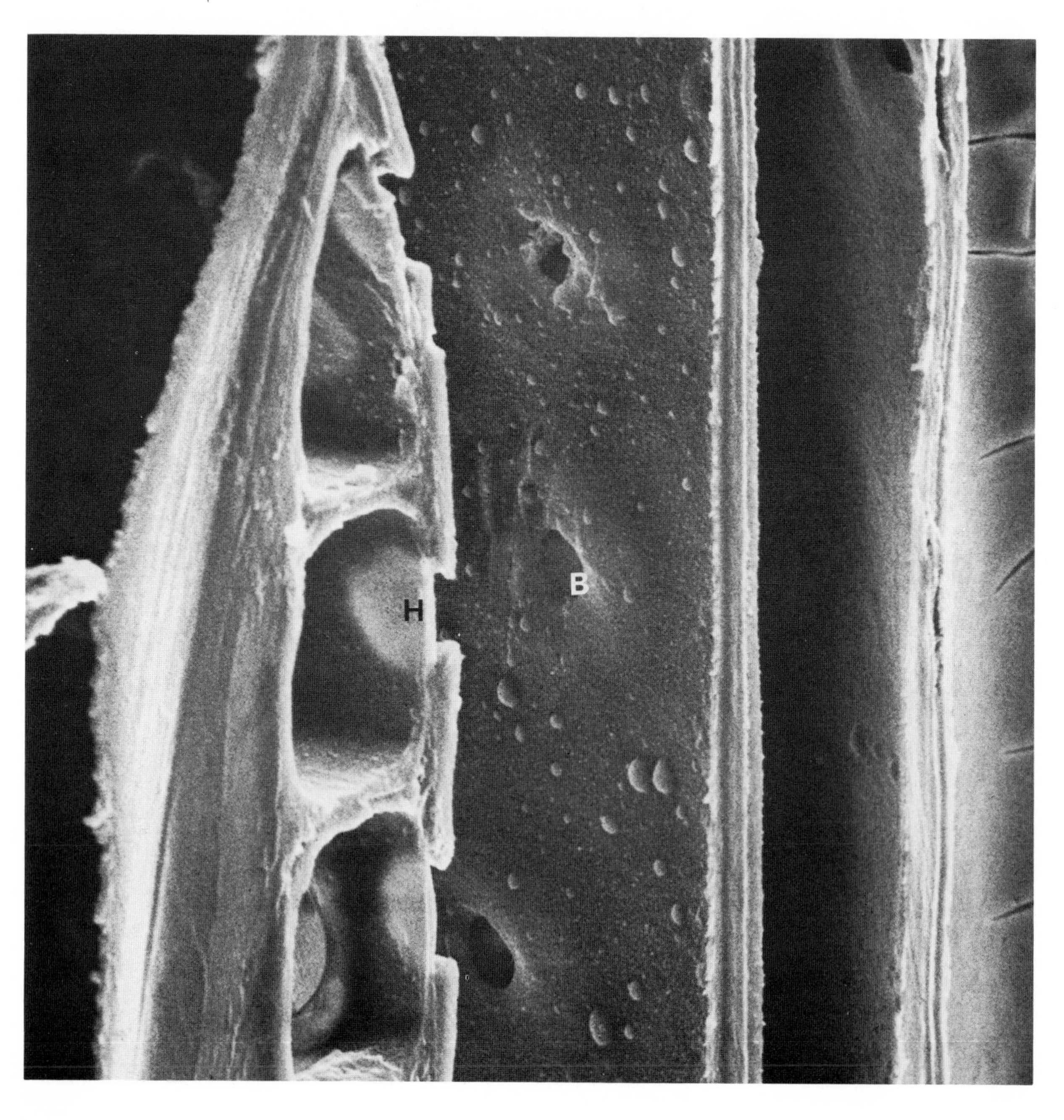

Sectional view of a bordered pit-pair in redwood
(Sequoia sempervirens)

Plate 84
X 6,000

This transverse section has passed through a bordered pit-pair. The border (**B**) surrounds the pit aperture which opens into the pit chamber (**C**). The pit membrane (**M**) is expanded in its central part (not clearly visible in the photo) to form the torus. Consult Esau (1965) for further details.

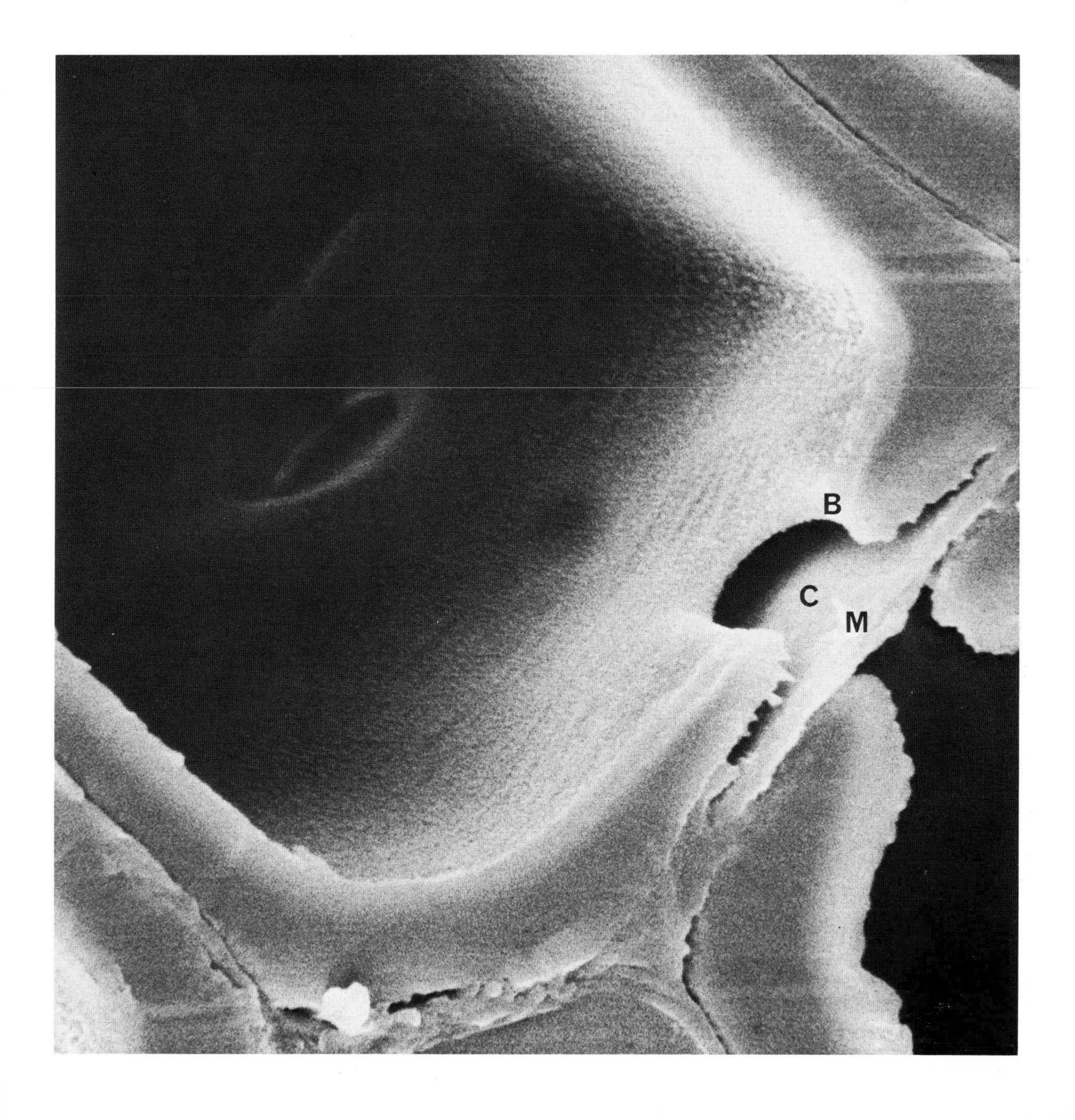

Plate 85

X 200

Spring and summer wood in the stem of redwood *(Sequoia sempervirens)*

The largest tracheids (lower right) are formed in the spring. At the end of the growing season, in summer, smaller tracheids are formed, with thicker walls (lower left to upper right). The smallest thick-walled cells in this region are intermediate in structure between tracheids and fibres. When growth begins again in the following spring, larger tracheids are formed (upper left). The distance between successive zones of smaller tracheids represents a year's growth and is termed an "annual ring". Annual rings show clearly in the wood of many temperate and cold climate trees, where there are definite growing and resting seasons. "False rings", which do not represent a full year's growth, can be formed when, for example, a sudden unusual drought in the middle of the growing season causes premature formation of "late" (summer) wood.

Redwood tracheids in transverse view *(Sequoia sempervirens)*

This is a more magnified view of some of the tracheids in the redwood stem illustrated in Plate 85. The middle lamella and primary walls (**W**) can be seen as a distinct region in the centre of the walls. They are surrounded by thicker secondary walls (**S**). The secondary wall is laid down inside the primary wall, usually at a time when cell elongation has ceased.

R—ray cells

Plate 86

X 900

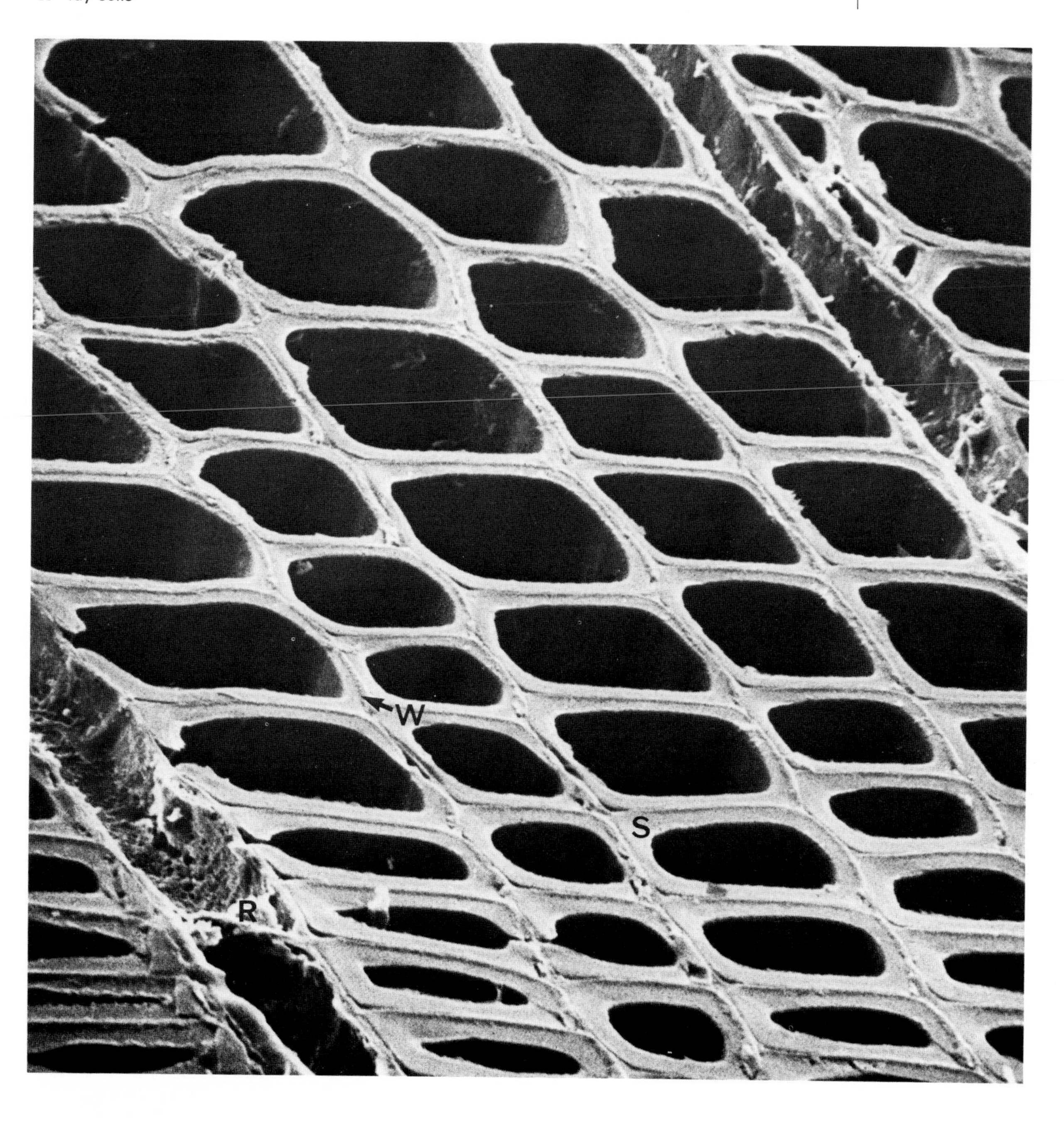

Plate 87
X 250

Surface view of a pine needle (*Pinus radiata* D. Don)

The stomata occur in rows on the needle leaves. The apertures which can be seen are not open stomata, but mark the openings to chambers above the stomata (see Plate 88).

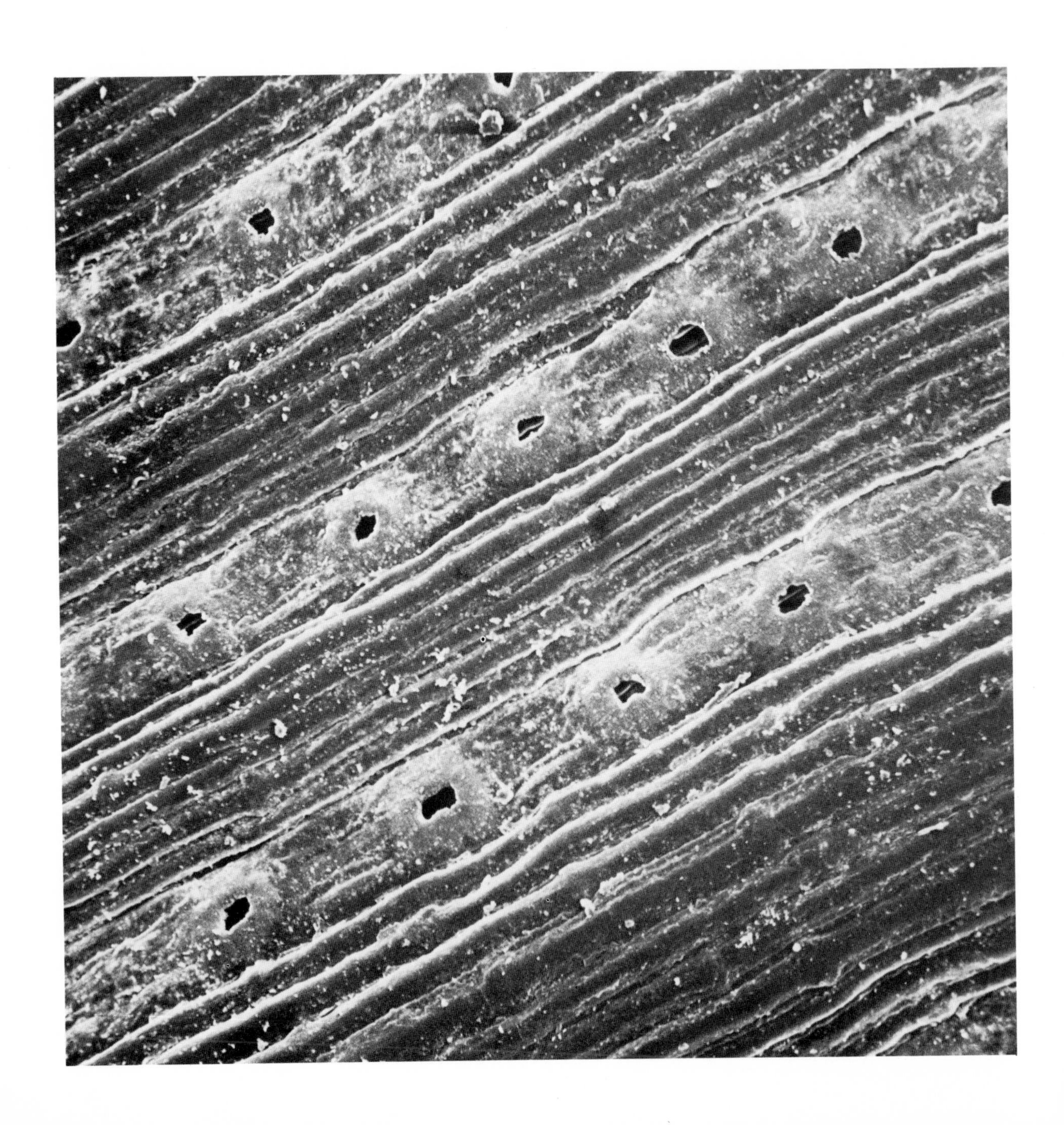

Transverse view of a stoma on a pine needle *(Pinus radiata)*

As illustrated in Plate 87 the guard cells of the pine-needle stoma are concealed in surface view by subsidiary and other epidermal cells. Subsidiary cells (**SB**) arch over the sunken stoma (**S**) forming a chamber which is lined with hair-like wax deposits on the surface of the cuticle. The lower central part of the chamber is formed by part of the guard cells and these exposed surfaces also bear the wax deposits.

G—guard cell

Plate 88
X 3,800

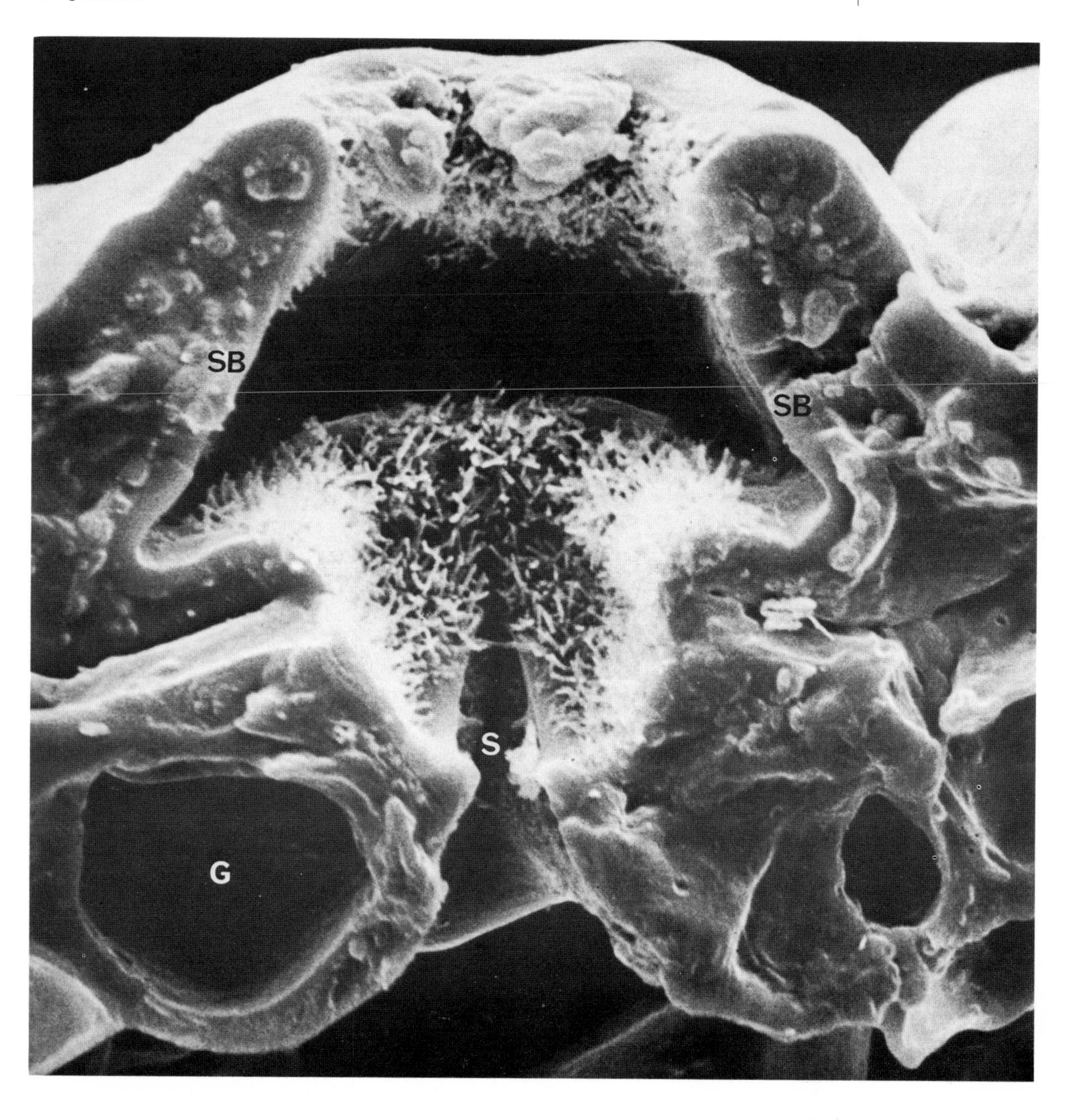

Plate 89
X 600

Transverse fracture of part of a pine needle *(Pinus radiata)*

Beneath the epidermis (**E**) is a hypodermis (**H**) of thick-walled fibre-like cells, one to several cells in thickness. They overlie the photosynthetic mesophyll tissue (**M**). The mesophyll cells have characteristic lobings and adjacent lobes are usually close to each other. It is possible that the lobes increase the surface area of the cells to facilitate gaseous exchange with intercellular air spaces. An endodermis (**D**), which is one cell thick, encircles the central tissue of the leaf. In the centre of this region are one or two vascular bundles. Phloem (**P**) and xylem (**X**) can be distinguished within the vascular bundle at lower right. For details of other cell types near the centre of the pine needle, which are not readily apparent in the photo, consult Esau (1965).

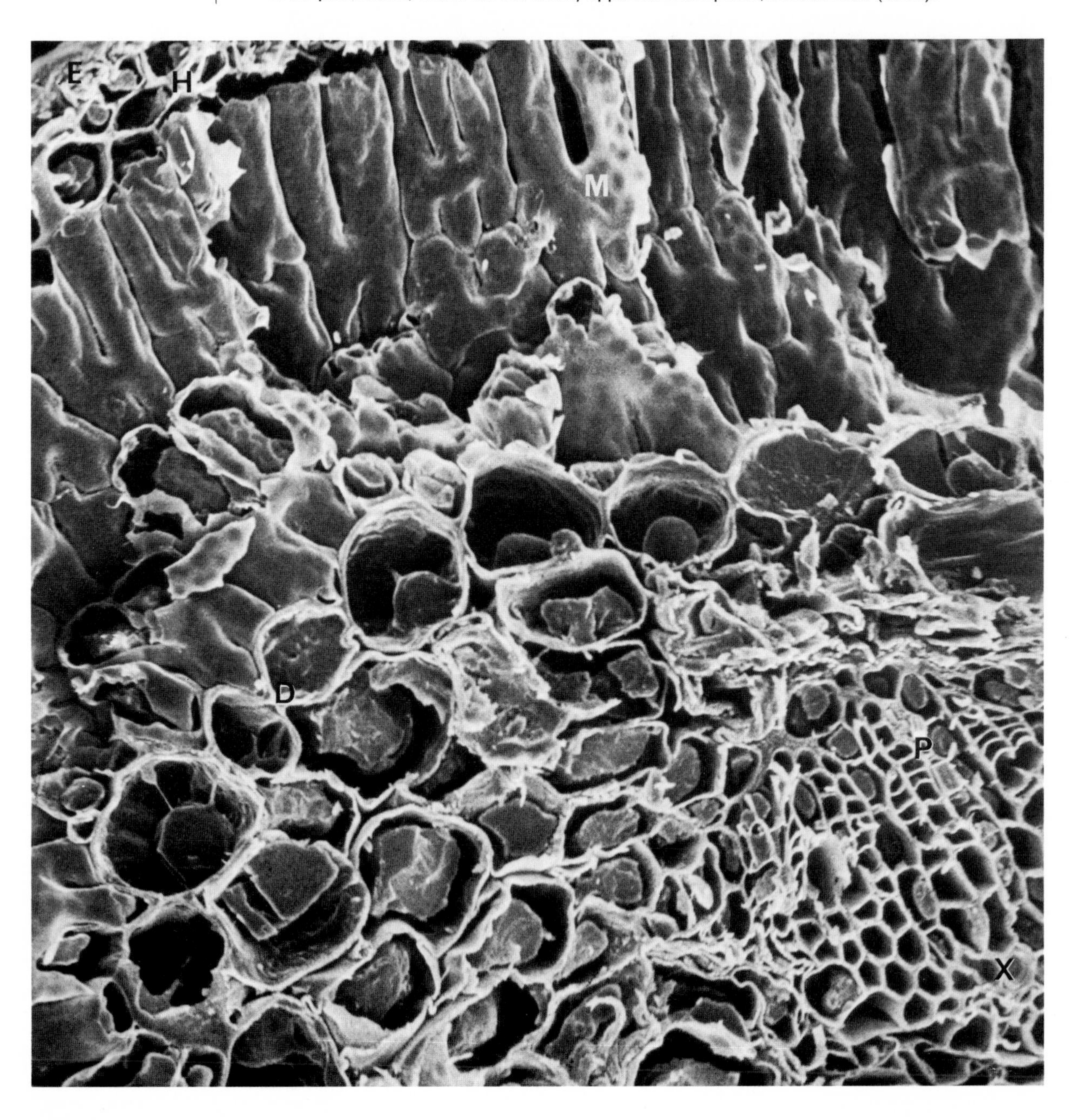

Male cone of Monterey pine *(Pinus radiata)*

Plate 90

X 50

Male cones are borne in a spiral cluster near the tip of a branch. Each cone replaces a short shoot, which includes the needles. The cone is formed in the axil of a scale leaf. A number of spirally arranged scale leaves are around the base of the cone above. When mature the cone is about 1 cm long. It consists of an axis bearing more than 100 microsporophylls. Two microsporangia (pollen sacs) are on each microsporophyll.

Plate 91
X 2,500

Pollen grain of Monterey pine *(Pinus radiata)*

The pine pollen grain (male gametophyte) has two lateral outgrowths, the air sacs. These give greater buoyancy to the wind-dispersed pollen. Each air sac is formed when the inner (endexine) and outer (ektexine) layers of the outer wall of the grain (exine) become separated in a particular region of the wall.

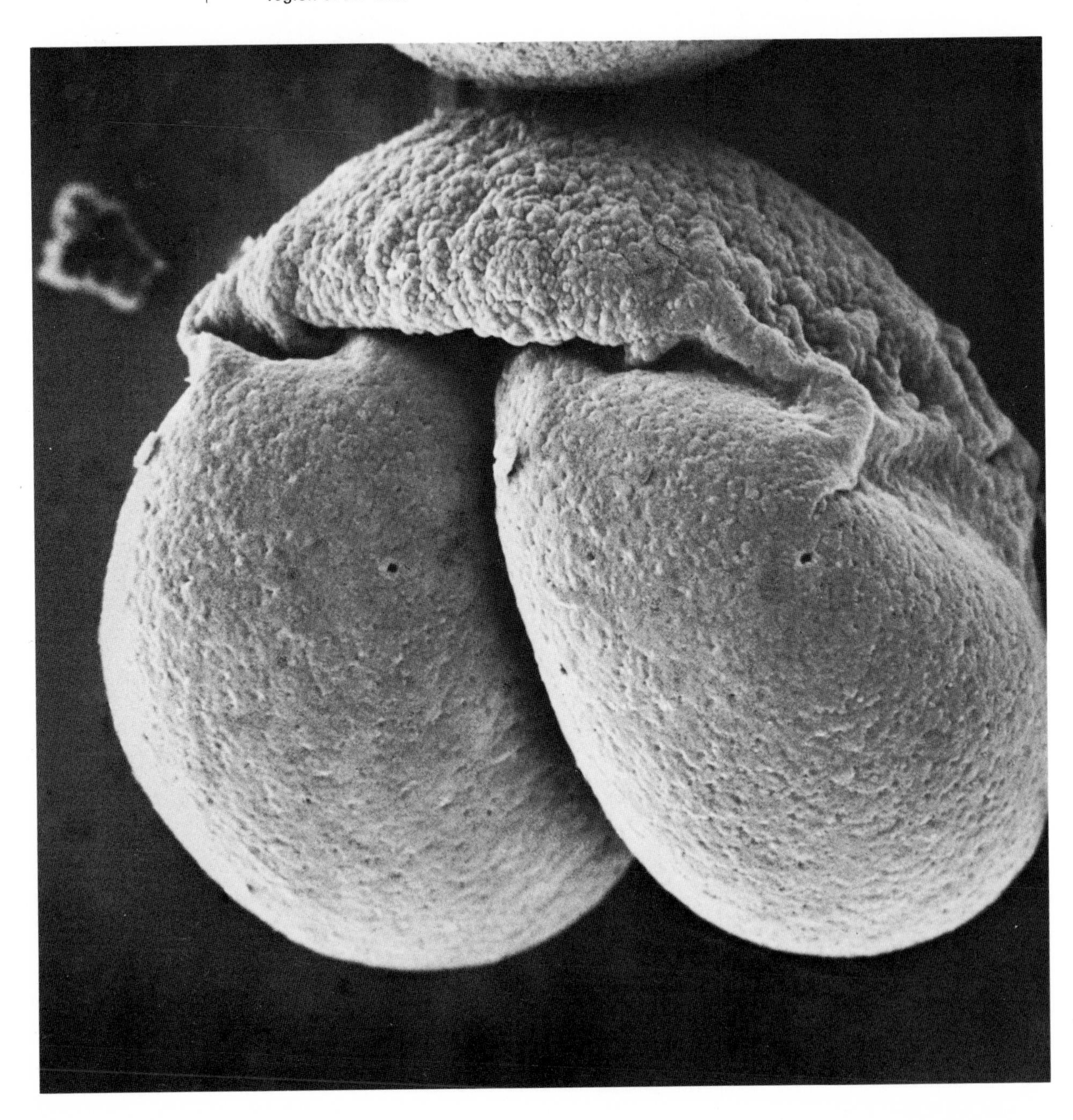

Pollen grain of Monterey cypress (*Cupressus macrocarpa* Hartw.)

Plate 92
X 5,700

The spheroidal grains of this Californian species are without apertures. *Cupressus* pollen is studded with tiny spherical particles (orbicules) which themselves bear minute spherical knobs. In contrast to the pollen of the pine family, which is, with a few exceptions, winged (Plate 91), the Cupressaceae has only non-winged pollen.

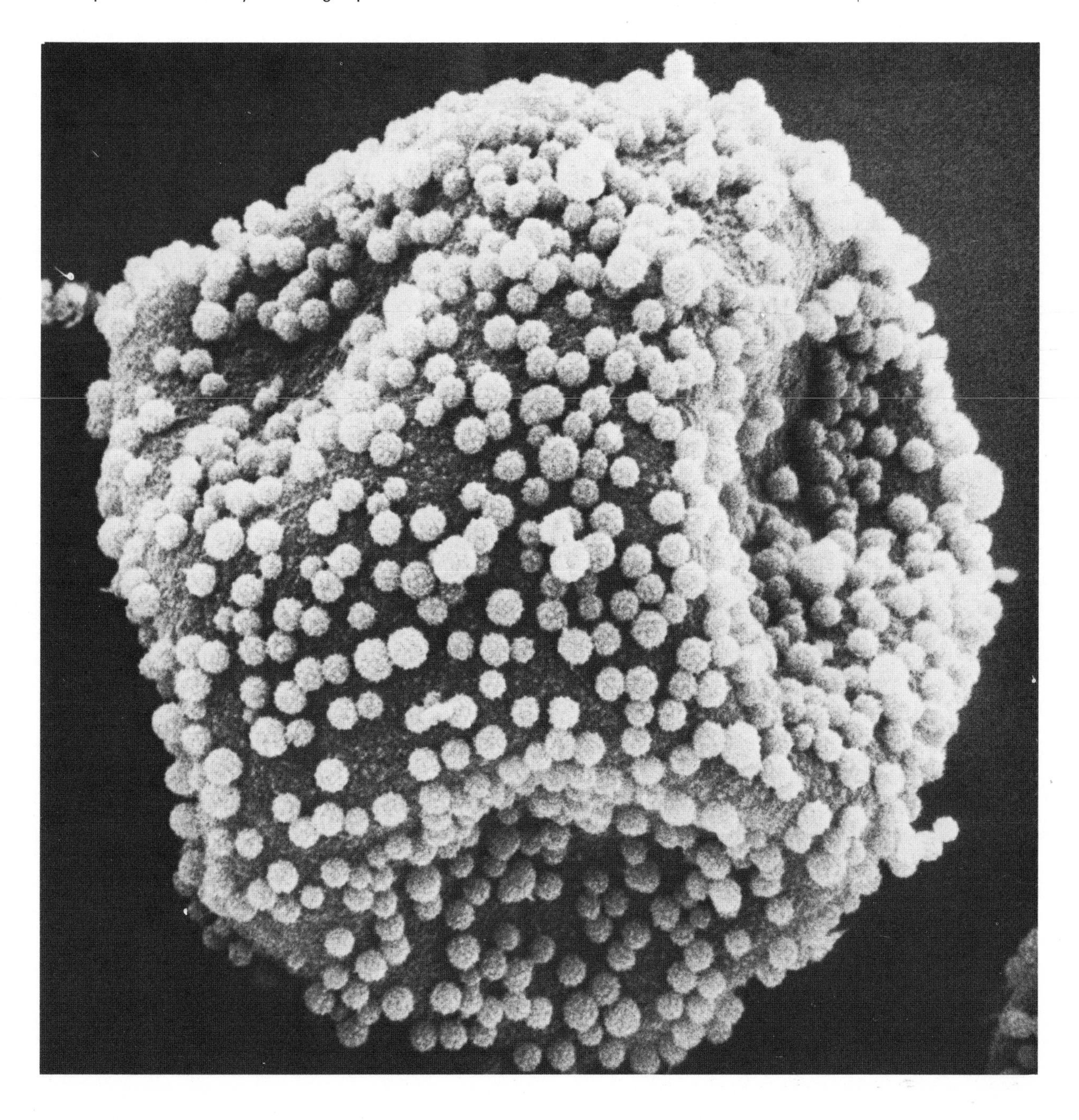

Plate 93
X 320

Ovule of pine (*Pinus muricata* D. Don)

The female pine cone consists of many spirally arranged cone scales, which are attached to a central axis. Each cone scale consists of two partly fused appendages—an ovuliferous scale and a bract scale beneath it. On the upper surface of each ovuliferous scale are two ovules, one of which is shown in the photo. The central tissue of the ovule is enclosed by a covering, the integument. The integument has a tubular opening, the micropyle (**M**), which faces towards the cone axis. Beyond the micropyle the integument has two prolonged horn-like processes.

At the time of pollination the cone scales are sufficiently spread apart to allow the wind-blown pollen grains to reach the micropyle of the ovules. They adhere to a sticky substance, the pollination drop, which is secreted by the ovule. As the drop dries the grains are drawn through the micropyle into a micropylar chamber, close to the developing female gametophyte. Fertilisation of an ovum by a male gamete from a pollen grain occurs approximately a year after pollination.

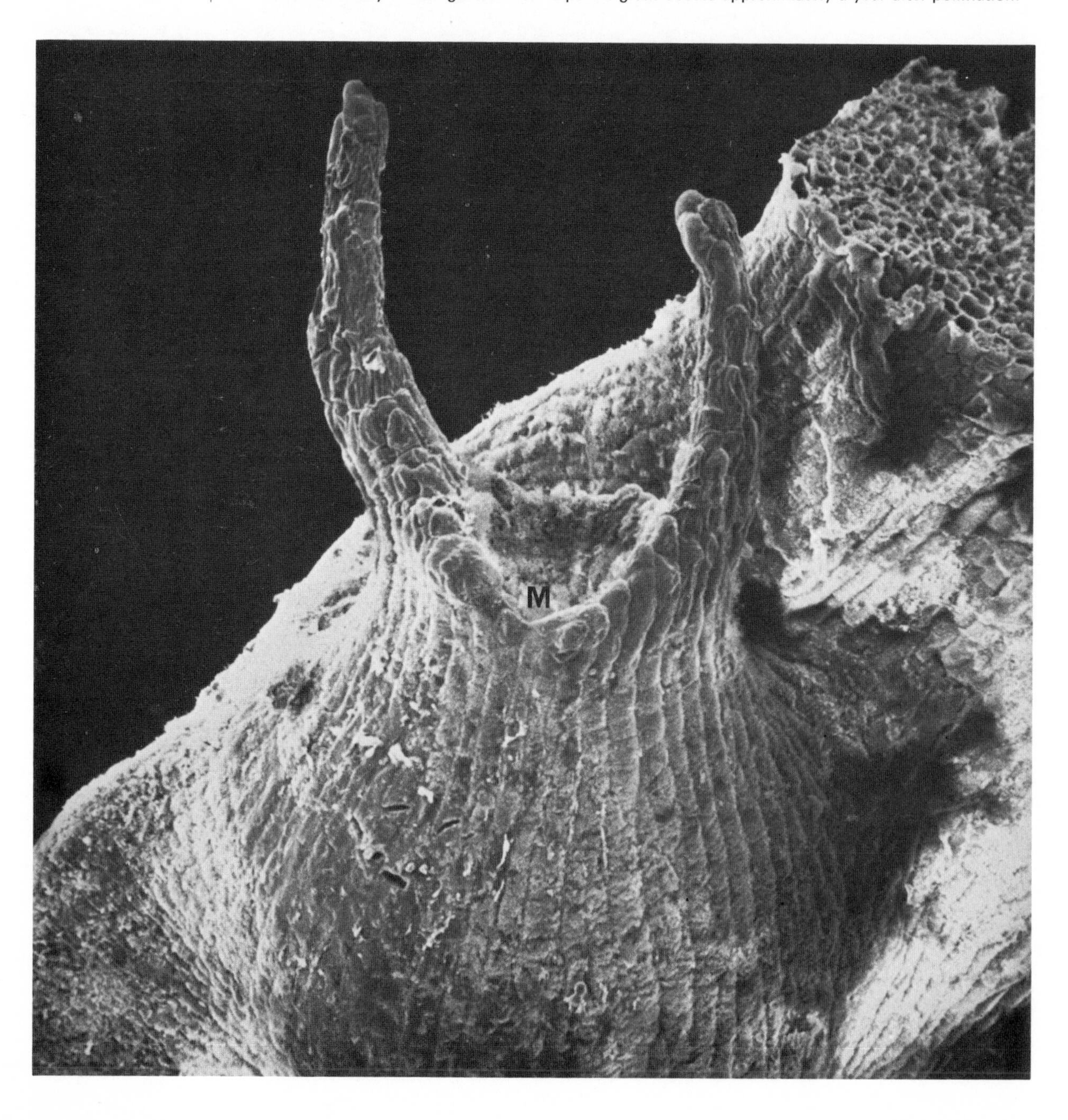

CLASS ANGIOSPERMAE

Flowering plants. These are seed plants in which the ovules are enclosed in an ovary. They are the only group of plants in which double fertilisation occurs.

Parenchyma cells in the petiole of celery (*Apium graveolens* L.)

Plate 94

X 4,600

Parenchyma cells are living, and generally thin walled, and form the relatively unspecialised "ground tissue" of plants. Much of the plant consists of parenchyma, including the cortex of root and shoot, all or most of the pith, and photosynthetic mesophyll tissue of leaves. Parenchyma also occurs on the xylem and phloem. The parenchyma cells below are from the cortex of celery petiole. They have a typical polyhedral shape and small intercellular air spaces (**S**) are present. It is tempting to suggest that the spherical structures which are appressed to the walls of some cells are nuclei. They are the same size as nuclei but it seems surprising that nuclei would survive the treatment for SEM examination. They may well be nuclei, however, as we were unable to find other cellular inclusions of the same size in an examination of fresh material under the light microscope.

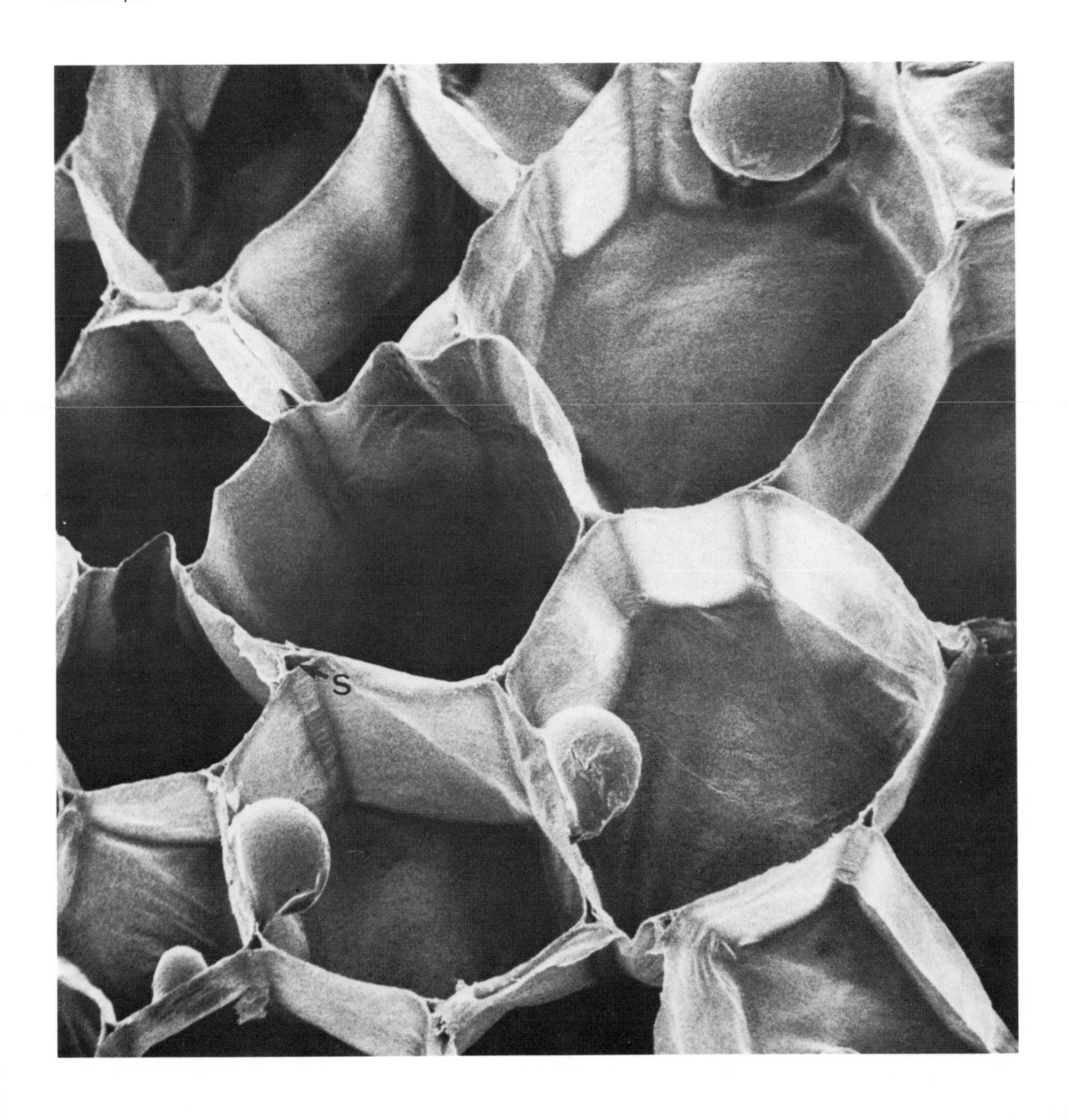

Plate 95
X 760

Stellate parenchyma in rush stem (*Juncus* sp.)

This illustrates an unusual type of parenchyma found in only a few plants. Each cell is star shaped, with a number of "arms" radiating from the centre of the cell. Each arm links with one from a neighbouring cell. Arrows indicate boundaries between cells. The system of large air spaces in the rush stem provides aeration for the plants, which grow in waterlogged soils. It has also been suggested that such a network provides considerable strength for a minimum amount of tissue and would protect the plant against the mechanical stresses and strains of an aquatic environment (Cutter, 1969).

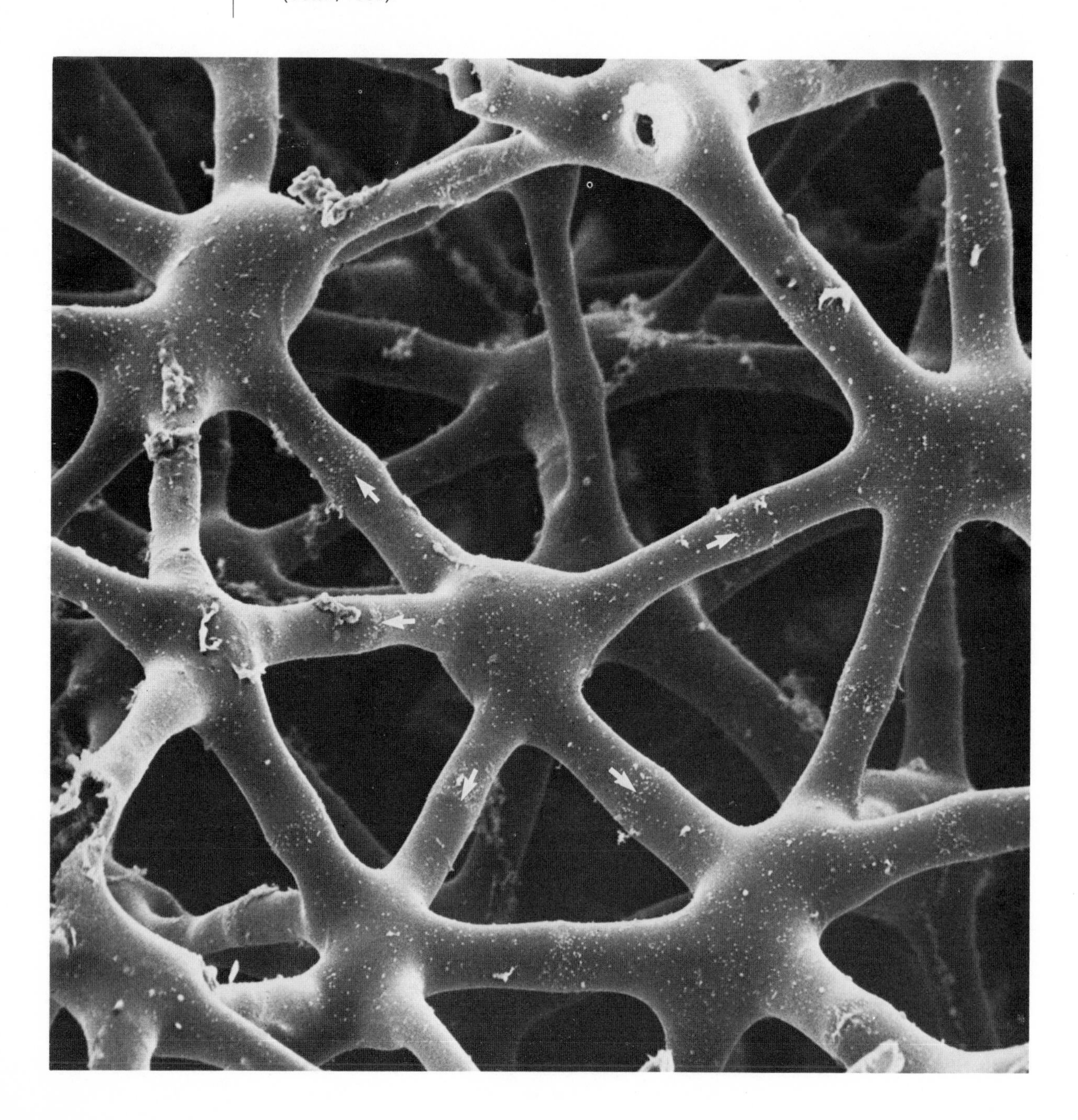

Collenchyma cells in the petiole of celery *(Apium graveolens)*

Plate 96
X 4,100

Collenchyma is living tissue in which the cells are usually elongated along the long axis of an organ and have thickened non-lignified primary walls. They lack secondary walls. Wall thickening is usually deposited unevenly and some parts of a wall are therefore thicker than others, as can be observed in the transverse fracture below. here the walls are thicker in the corners (angles) of the cells, forming what is termed angular collenchyma. Collenchyma is thus living tissue with a supporting function. It gives strength to young organs and to herbaceous plants. It is usually situated near the outside of an organ and is common in stems and petioles but rare in roots. In the leaf stalk of celery it occurs in patches beneath the epidermis.

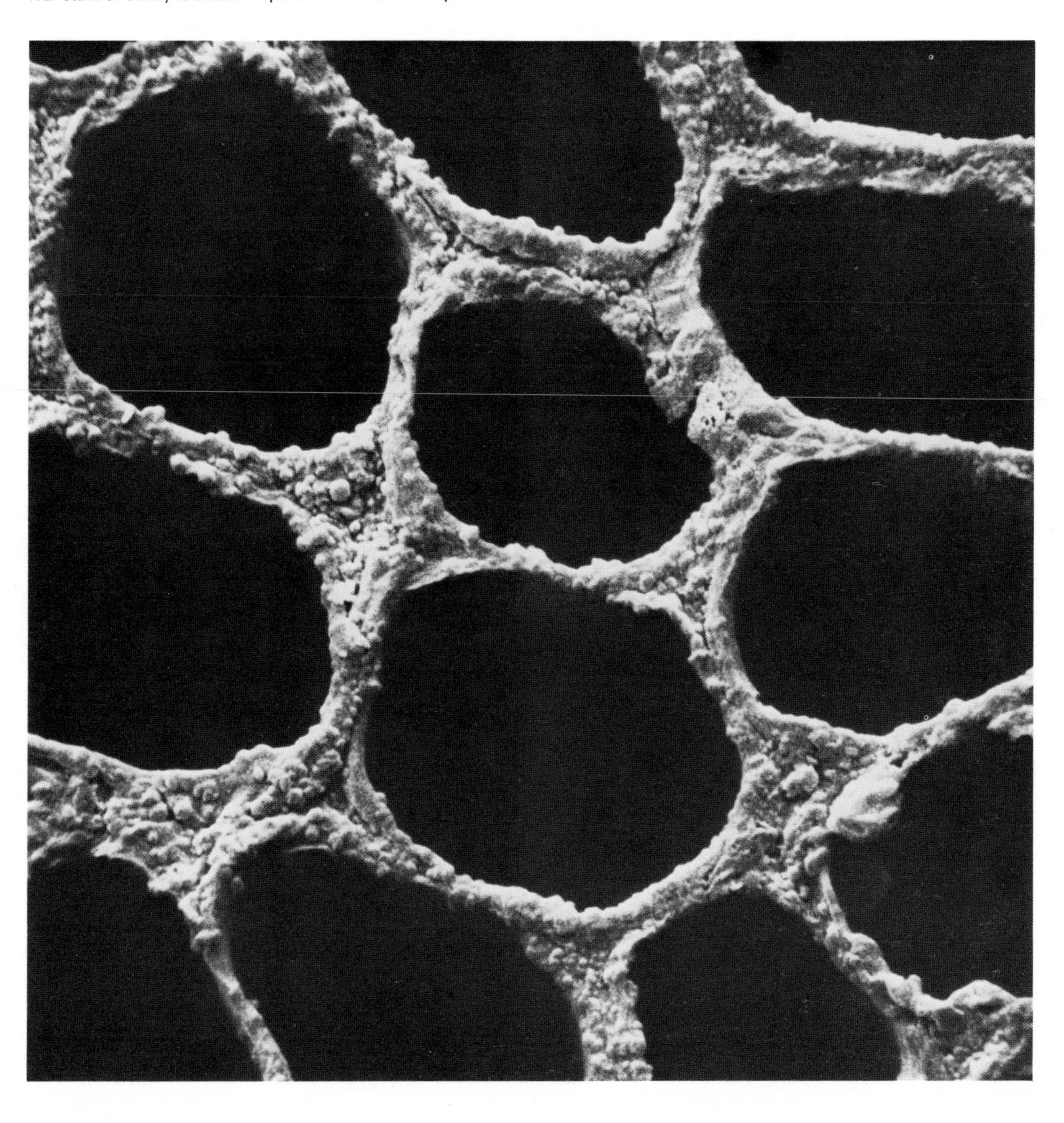

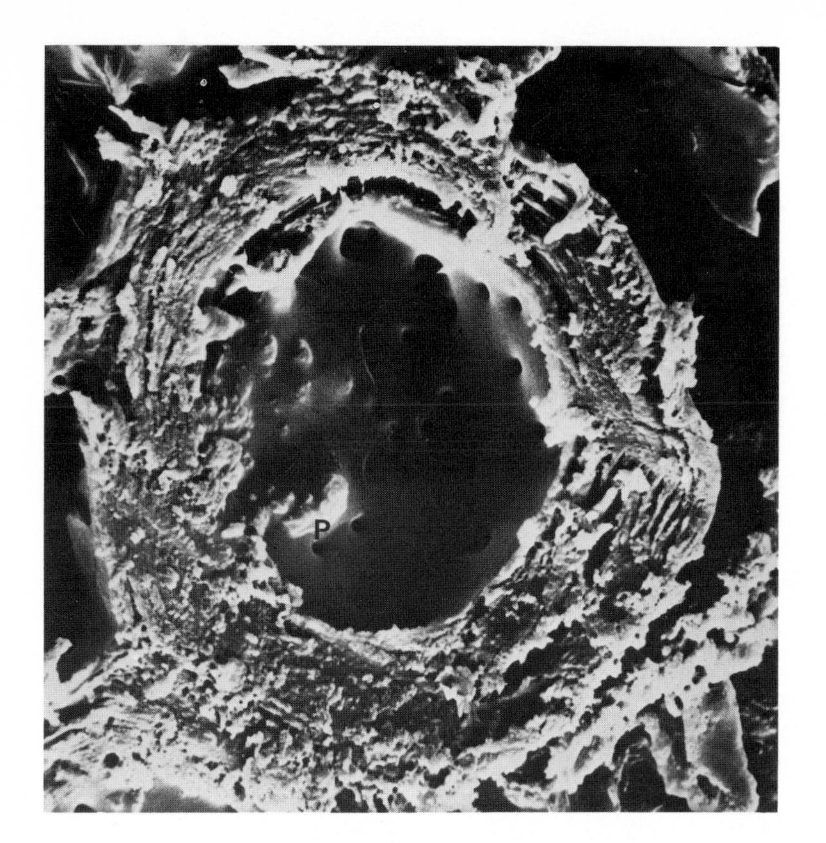

Plate 97 X 1,500

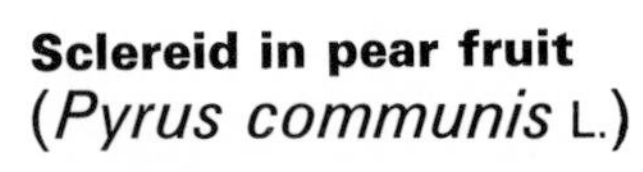

Sclereid in pear fruit (*Pyrus communis* L.)

Sclereids and fibres (Plate 98) are classified as sclerenchyma tissue. They have thick secondary walls which are normally lignified and are usually without protoplasm when mature. Their principal function is mechanical support. Sclereids are variable in shape, but in general one diameter is not more than three times the size of another.

Sclereids give the grittiness to the flesh of pears. This type of sclereid, which has the same shape as parenchyma cells, is termed a brachysclereid or stone cell. Note the thick wall and the pit apertures (**P**) on the inside of the wall in this transverse section.

Plate 98 X 2,000

Fibres in New Zealand flax leaf (*Phormium tenax* J. R. et G. Forst.)

In contrast to sclereids (Plate 97), fibres are elongated cells. The thick lignified walls are shown in transverse section. The fibres are mature and without protoplasm.

Fibres occupy a considerable proportion of the flax leaf and the plant is used in industry for the manufacture of hard-wearing floor coverings.

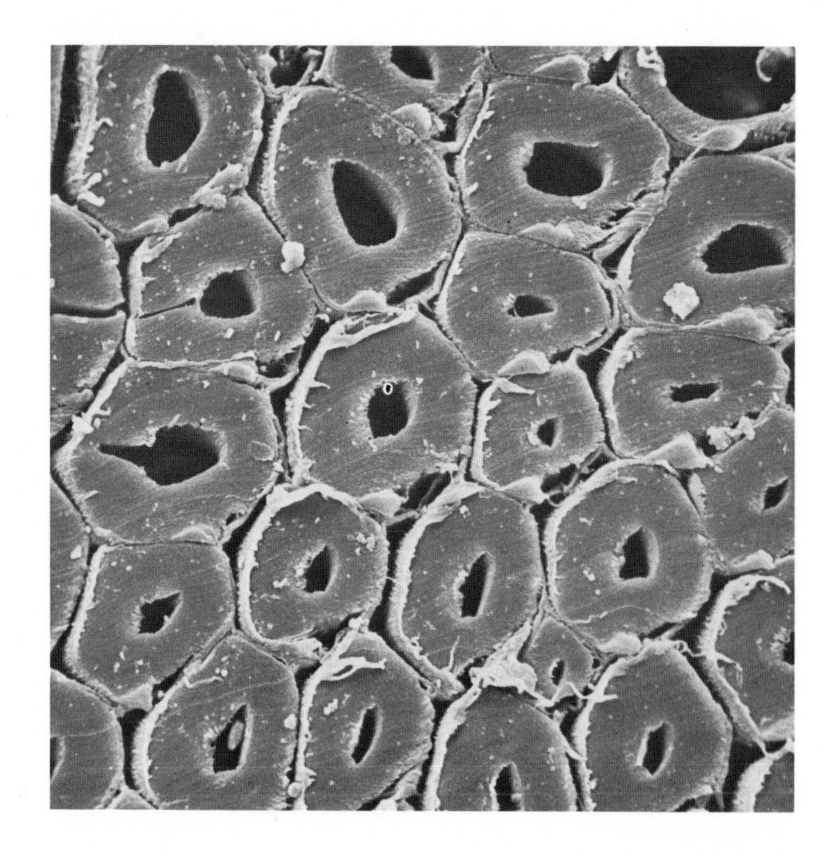

Cork cells of *Quercus suber* L.

Plate 99
X 3,300

A cork from a bottle was sectioned for this photo. Commercial cork is obtained from a Mediterranean species of oak.

Cork cells are dead at maturity and their walls are impregnated with a waterproof substance, suberin. Cork cells of *Quercus* are filled with air and form a light, waterproof, compressible but resilient tissue.

The term "cell" was coined by Robert Hooke in 1664 to describe cork tissue as seen with a microscope (see Weier, Stocking and Barbour, 1970).

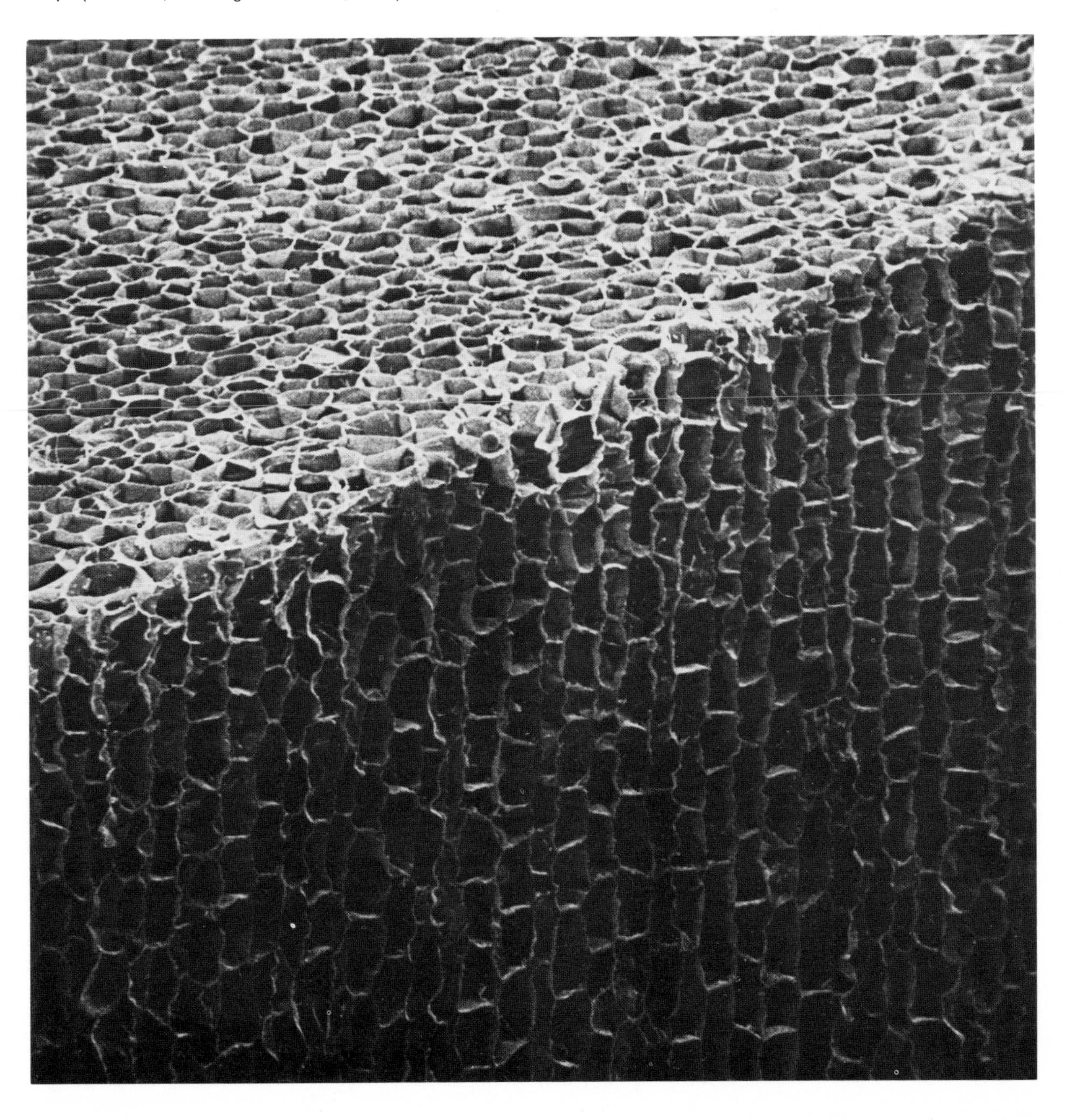

Plate 100 X 5,500

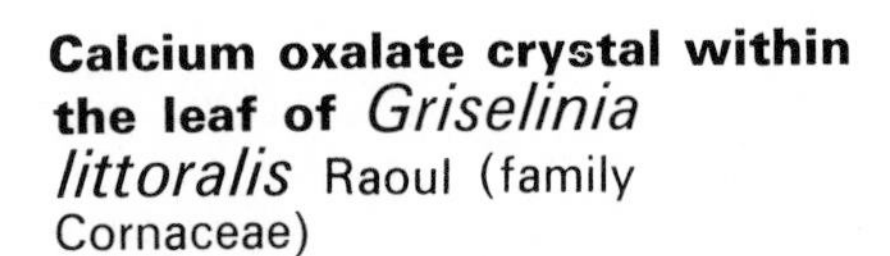

Calcium oxalate crystal within the leaf of *Griselinia littoralis* Raoul (family Cornaceae)

Crystals are deposited in the cells of many plants. They are believed to be deposits of waste products and calcium oxalate crystals are the most common. The form of this crystal is a druse, which consists of a more or less spherical cluster of many pyramid-shaped crystals with projecting points.

The intricate manner in which calcium oxalate crystals grow within the vacuoles of some plant cells, as seen under the electron microscope, has been described by Ledbetter and Porter (1970).

Plate 101 X 700

Lithocyst in a leaf of the india-rubber plant (*Ficus elastica* Roxb.)

Lithocysts are large specialised epidermal cells, each containing a cystolith. The cystolith (**C**) is formed when a stalk-like ingrowth of the cell wall becomes impregnated with calcium carbonate. The nucleus apparently remains in a functional condition within the lithocyst (Cutter, 1969). It is considered that the calcium carbonate is a waste product.

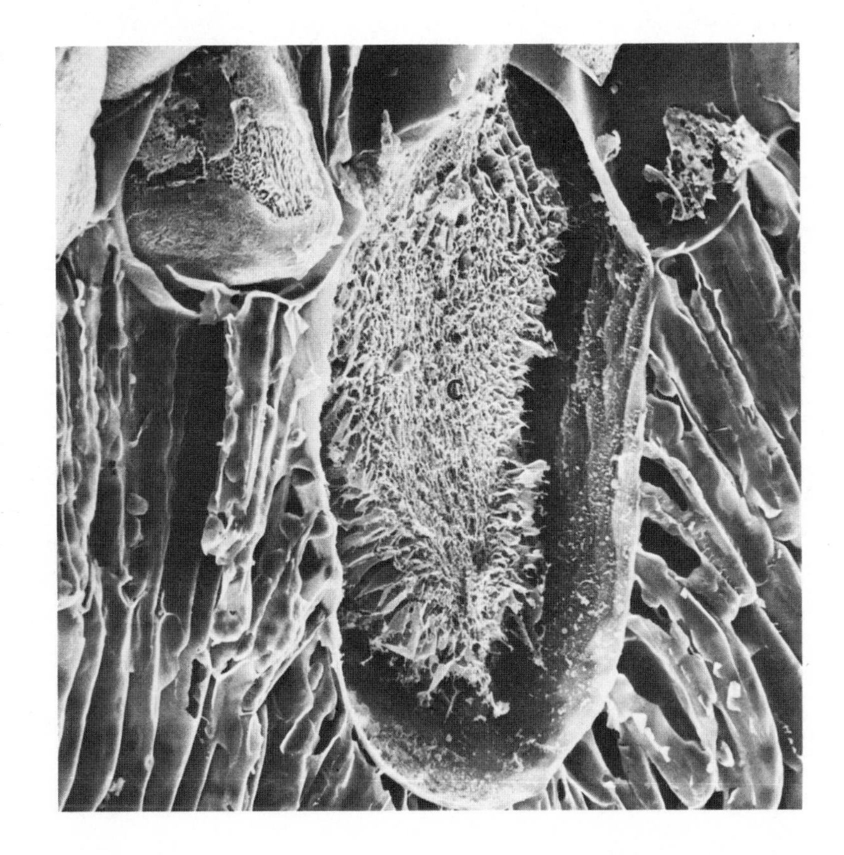

Tyloses in a vessel of a corn stem (*Zea mays* L.)

Plate 102

X 500

The two balloon-like structures in this large vessel are tyloses. In the lower part of the photo can be seen two tyloses which collapsed when the material was sectioned. Each tylosis is formed when a parenchyma cell grows out through a pit-pair connecting it with the vessel. The pit membrane is ruptured during this growth. The nucleus of the parenchyma cell moves into the tylosis, together with some of the cytoplasm. Tyloses may remain thin walled but in many plants they develop lignified secondary walls. Tyloses appear in vessels when they become inactive or when the xylem tissue has become injured. They are very abundant in certain plants, e.g. grape, entirely absent in others and have been reported in a number of herbs.

Many durable woods contain tyloses, which, because they block the vessel lumina, prevent the entrance of water, air and fungal filaments. This is a disadvantage in the timber industry. for it is difficult for preservative to penetrate such wood.

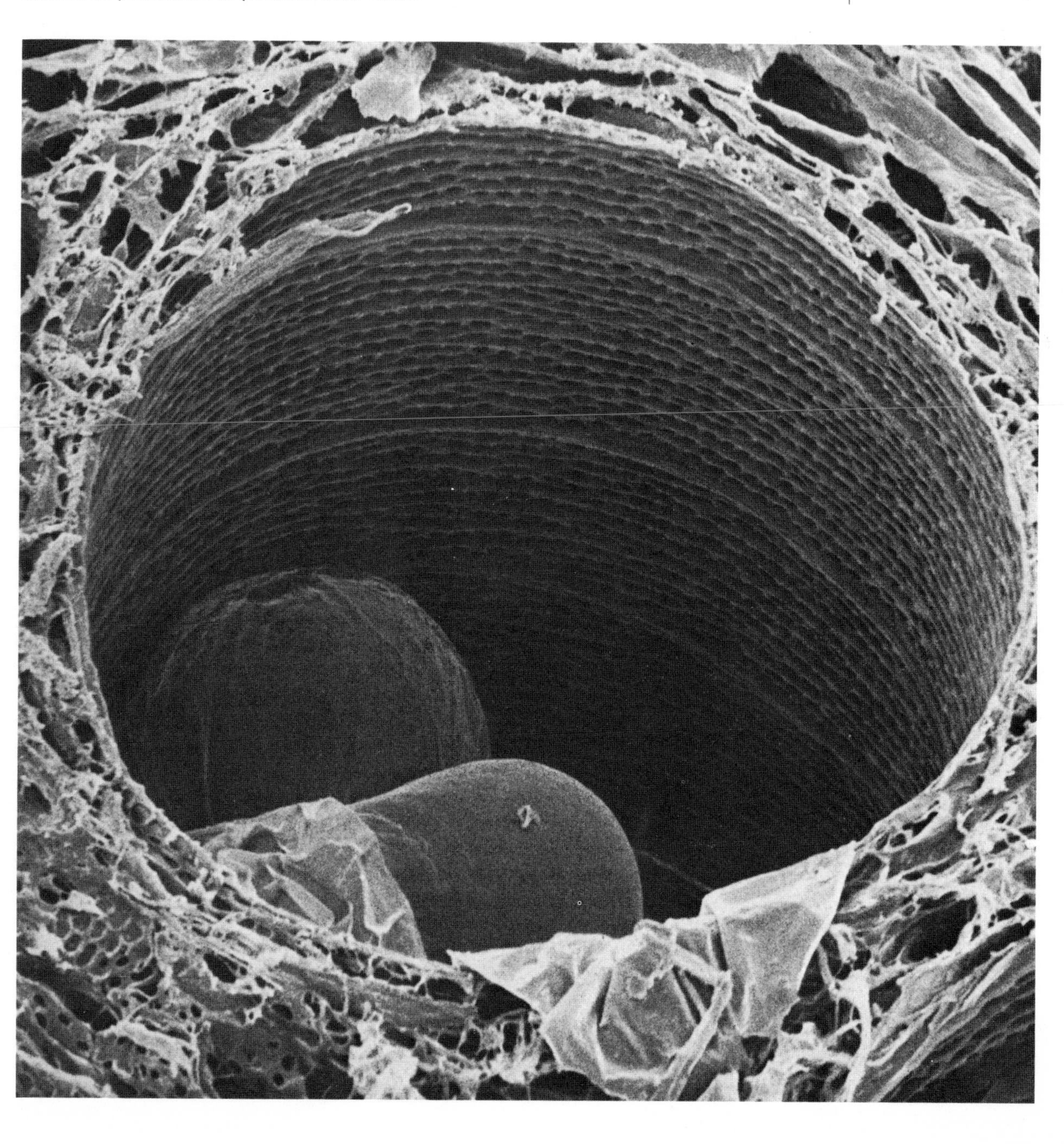

Plate 103
X 630

Primitive vessels in the stem of *Eupomatia laurina* R. Br.

The long oblique end walls, shown in the side view (Plate 103, centre) and face view (Plate 104), contain up to 100 or more scalariform perforations. In the upper part of the end wall of a vessel in Plate 104 some scalariform pits (**P**) are present, which differ from perforations in having a membrane. Note that the pits on a side wall of this vessel element, and others, are small and rounder. Transitions between pits and perforations can be found near the tips of the end walls, where there are incomplete pit membranes (Plate 105). The end wall has been cut and the full width is not present. The upper pit membrane is almost complete. The bordered nature of the

pits and perforations can be seen. The borders are not as extensive as those of conifer tracheids (Plate 83).

Both tracheids and vessel members conduct water in the xylem. They are dead at maturity and have variously thickened walls. Whereas tracheids have pits in their end walls, vessel members have perforations. A pit represents a thinner part of a secondary wall—thin enough and with a microfibrillar mesh sufficiently coarse to allow water to pass comparatively readily from one cell to another—but a perforation is a hole in a wall. Vessels have evolved from tracheids and stages

Plate 104

X 870

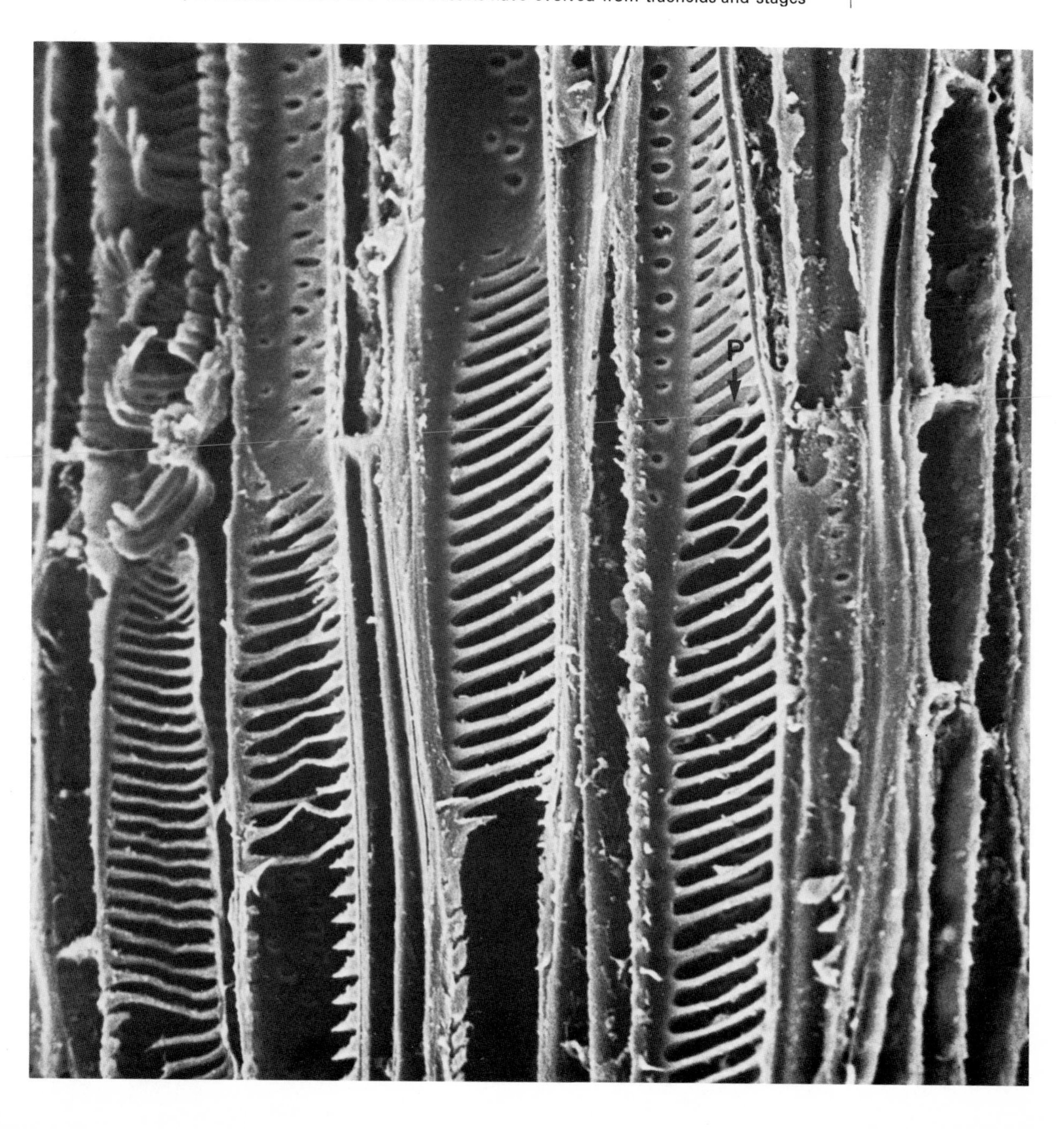

Plate 105

X 8,400

in their specialisation are well preserved among living angiosperms. A vessel is a vertical series of vessel members, one above the other, which have perforations in their end walls. Vessel members evolved in angiosperms from long tracheids of small diameter, with long obliquely sloping end walls, containing numerous scalariform bordered pits. The most primitive vessel members resemble these tracheids except that the pit membranes have disappeared from some of the pits on the end walls. *Eupomatia*, a genus of two species found in Australia and New Guinea, has vessels which are probably the most primitive in the angiosperms (Eames, 1961). It is noteworthy that the family Eupomatiaceae is part of the Magnolian alliance (Annonales), which includes the most primitive angiosperms, some of which lack vessels.

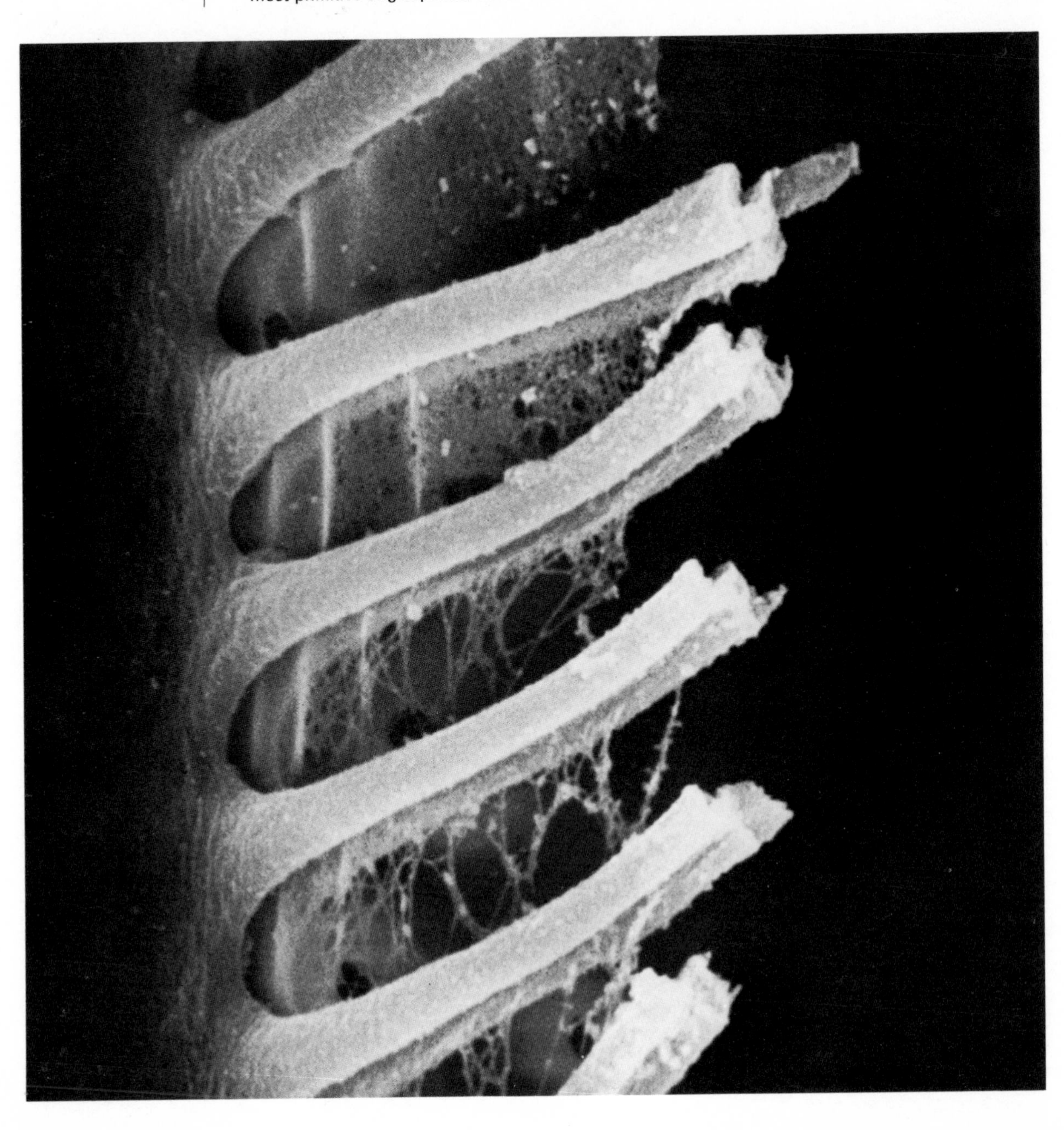

Plate 106 X 1,100

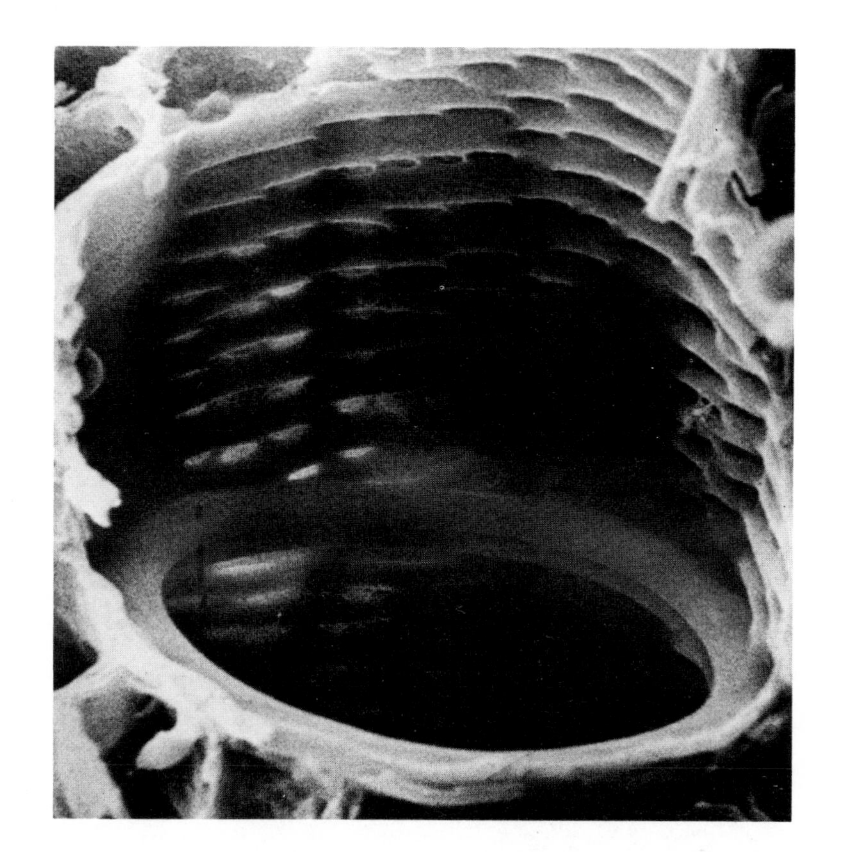

End wall of advanced vessel member in corn stem
(*Zea mays* L.)

There has been considerable work done, including statistical analyses, on the phylogenetic specialisation of vessels. During their evolution from the primitive type (Plates 103–105) vessel members became shorter, wider, rounder (less angular in cross-section) and their end walls became less oblique and finally transverse, as in the photo. The perforations in the end walls also underwent evolution. The borders ceased to develop, then walls between individual perforations disappeared and in the most advanced types a simple perforation plate was formed, as shown at right. Here most of the end wall has become a single large perforation. Vessels have evolved independently from tracheids in three gymnospermous plants (*Gnetum, Ephedra* and *Welwitschia*) and in a number of pteridophytes (some ferns, e.g. *Pteridium* (Plate 69), *Selaginella* and *Equisetum*). It is also considered that vessels arose independently in the monocotyledons and dicotyledons (Esau, 1965). Vessels form a more efficient water-conducting system than tracheids and some of the success of the angiosperms has been attributed to this.

Plate 107 X 1,700

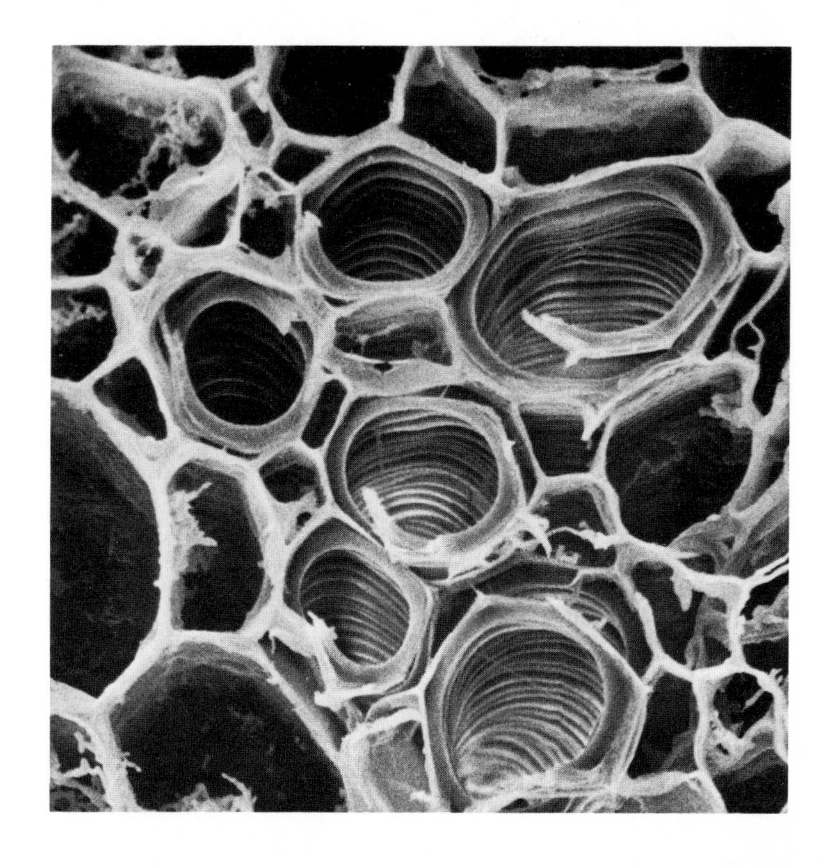

Protoxylem vessels in cotton
(*Gossypium* L.)

A group of spirally thickened protoxylem vessels are shown in this transverse view of a petiole of cotton. The protoxylem is the first-formed primary xylem and differentiates from elongated procambial cells, which originate at the apical meristems of the root and shoot. The protoxylem vessels have a continuous thin primary wall, with rings or spirals of secondary-wall material deposited inside it. They are extensible and during the elongation (maturation) of an organ, the rings or spirals become more widely separated. The extra thickening, therefore, gives extra strength to the vessels without greatly inhibiting their extensibility. The primary walls between the thickening are able to undergo considerable stretching before they are torn.

Plate 108 X 4,500

Sieve plate in the phloem of pumpkin stem *(Cucurbita pepo)*

Phloem is the main tissue involved in the transport of carbon compounds in vascular plants. In angiosperms, transport, primarily as sucrose, is in the sieve tubes. Other groups of vascular plants have less specialised sieve cells in the phloem.

A sieve tube is a longitudinal file of elongated sieve-tube members. In the most specialised forms, including pumpkin, the end walls between sieve-tube members consist of a single sieve plate. The sieve plate shown here has sieve pores (**P**) enclosed by a cylinder of callose (**C**). Between these is the cellulosic cell wall (**W**). The sieve tubes normally contain a proteinaceous material which may be in a filamentous form. When phloem is sectioned or subjected to other forms of injury, this proteinaceous substance, called slime, accumulates around the sieve plates. To obtain this photo we treated the section with a dilute solution of sodium hydroxide to remove the slime. This would have dissolved some of the callose around the periphery of the sieve pores.

Materials move from one sieve-tube member to another via the pores in the sieve plate. The mechanism by which carbon compounds are transported in the phloem is still a matter of controversy—for a discussion of the various mechanisms which have been proposed, see MacRobbie (1971). Transport of materials in the sieve tubes requires the expenditure of energy and the companion cells (**CC**) seem to be the source of at least some of this energy.

A sieve-tube member originates from the same mother cell as a smaller parenchymatous companion cell. Although mature sieve-tube members have lost their nuclei, they are living cells. The companion cells are closely associated with the sieve-tube members. The two types of cells are interconnected by means of plasmodesmata. There is a close physiological relationship too between them, and when a sieve tube ceases activity the associated companion cells also die.

For further details of the phloem in various plant groups, the reader should consult the comprehensive and well-illustrated work by Esau (1969).

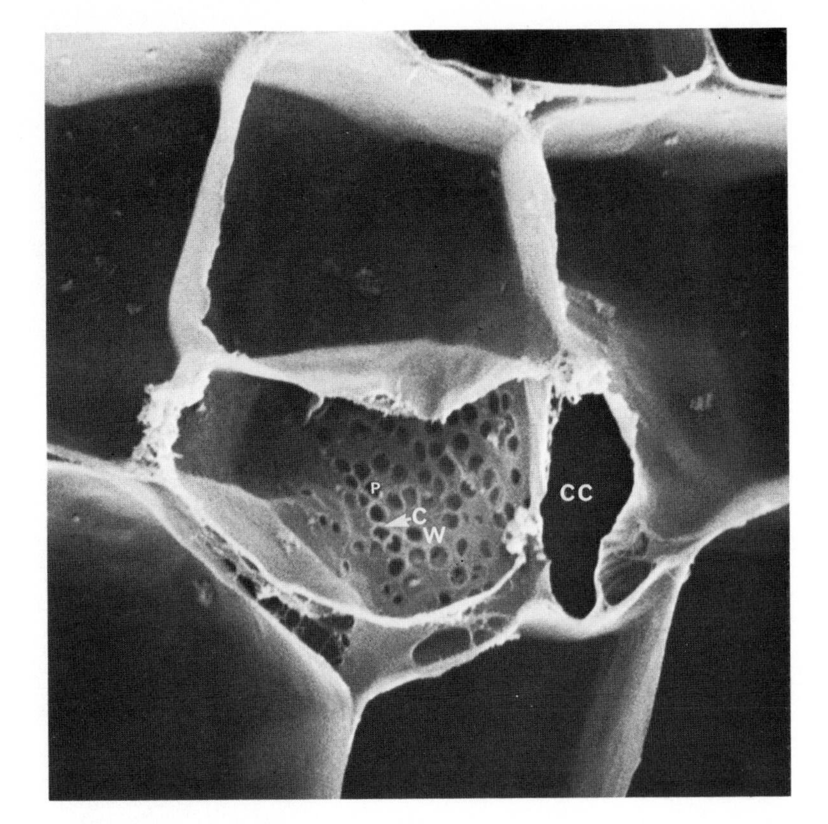

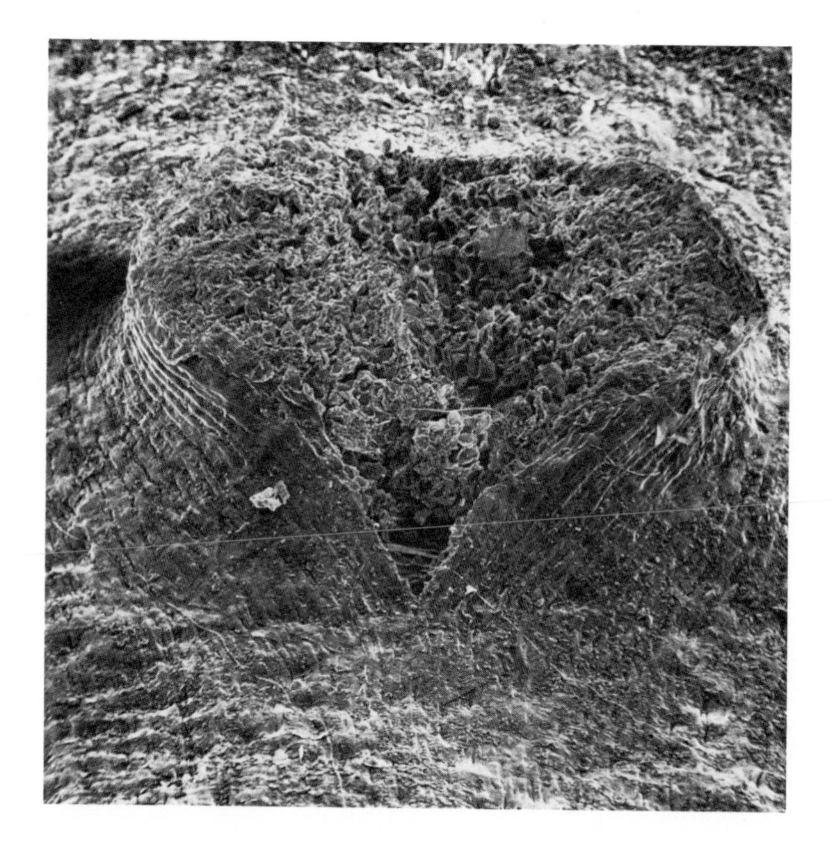

Plate 109

X 170

Lenticel on the stem of *Prunus* L.

In young woody stems gaseous exchange occurs between the atmosphere and the internal cells through the stomata. When cork is formed beneath the epidermis, this impervious layer prevents the exchange of gases. Cork is formed by the action of a cork cambium. The cork cambium is a ring of meristematic cells, formed from parenchyma in the outer cortex, which forms cork cells to the outside and parenchyma of the secondary cortex on the inner side. Lenticels are formed to enable gaseous exchange to continue between the atmosphere and the cells beneath the cork. In the lenticel region, the cork cambium produces thin-walled cells instead of cork. These thin-walled cells (complementary cells) have intercellular spaces between them. This loosely arranged tissue breaks through the epidermis and the lenticel projects above the stem surface, as shown.

Lenticels are generally present on the older roots and stems of woody plants.

Plate 110

X 260

Transverse section of a lenticel on *Prunus* stem

The exposed cells on the outside of the lenticel (top centre) weather away and are replaced by others developing from the cork cambium. In *Prunus* the loose complementary tissue alternates with more compact cells containing suberin, the waterproofing material which occurs in the walls of cork cells. These layers are called closing layers (**CL**) because they hold the loose tissue together. The closing layers are successively broken as further growth occurs from the cork cambium. The cork cambium beneath the lenticels also forms some secondary cortex (**2L**).

Alongside the lenticel can be seen a thick cuticle (**C**) on the epidermis, with cork tissue beneath. The cambial zone (**CZ**) is continuous with the cambial zone within the lenticel (**LZ**) and beneath it are a few layers of the secondary cortex (**2C**).

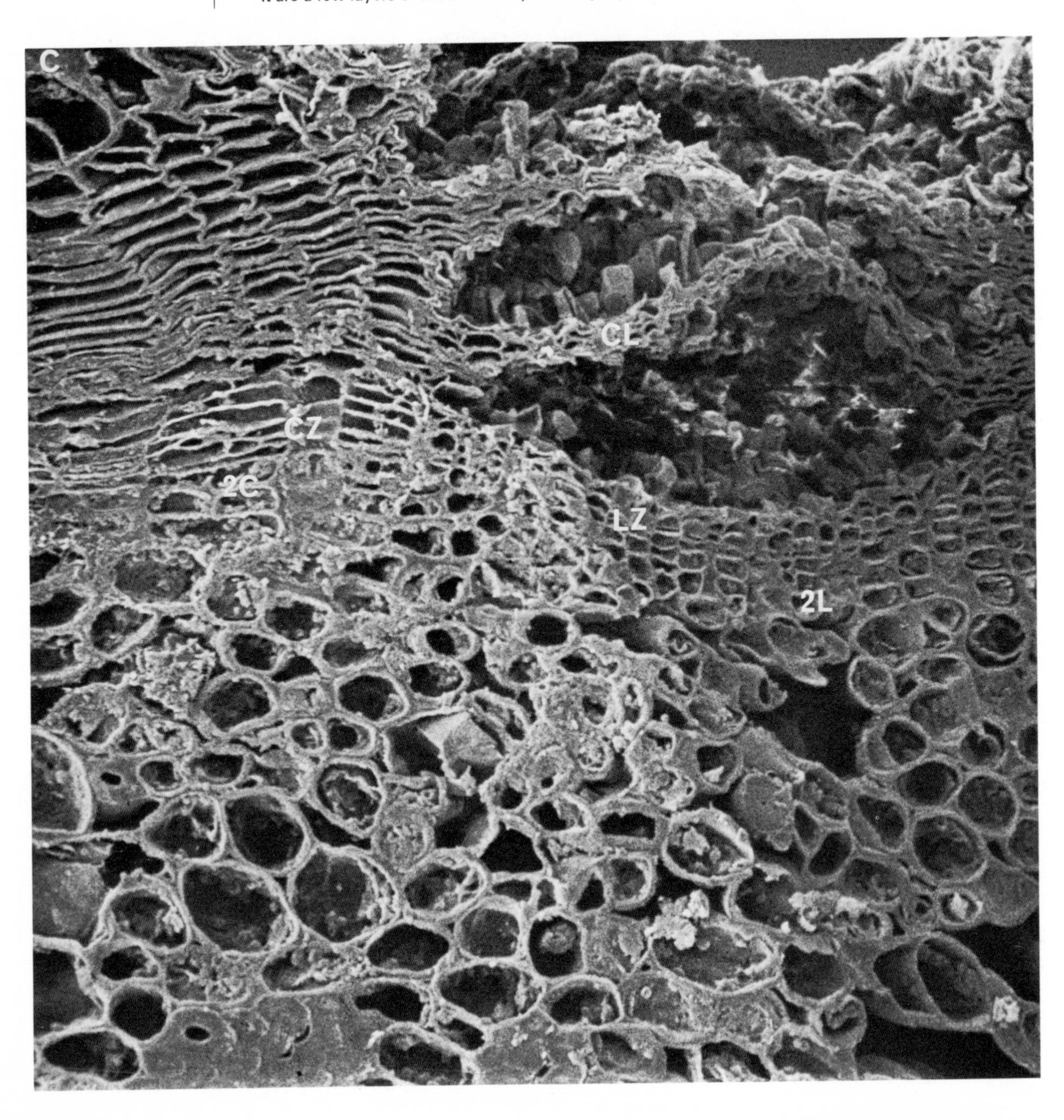

Stem apex of *Coleus* Lour.

Plate 111

X 170

At the tips of shoots and roots there are embryonic tissues—the apical meristems—where cell production continues throughout the life of the plant. The young leaves have been removed from an apical bud on a *Coleus* stem to reveal the youngest pair of leaves. Between them is the dome-shaped shoot apex. The next pair of leaves will arise on the flanks of the apical meristem at right angles to the present pair.

There are two types of epidermal hairs (trichomes) on the leaves; on the edge and outer (abaxial) face of each leaf are long pointed trichomes consisting of several cells arranged end to end. Smaller glandular trichomes on a short stalk one cell high terminate in an expanded head of up to four cells and occur on both faces of the leaves.

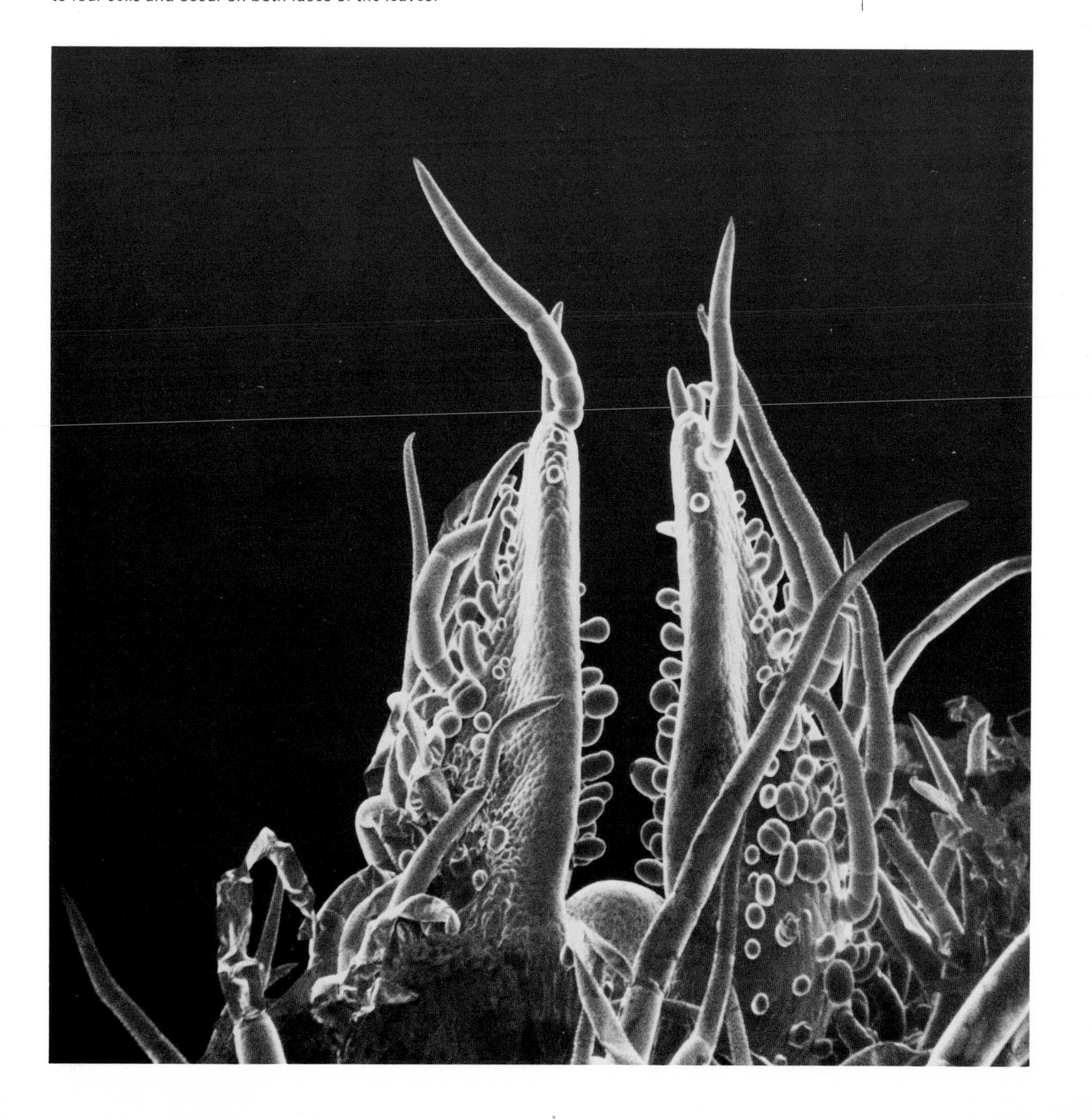

Plate 112
X 100

Vascular bundle in pumpkin stem (*Cucurbita pepo* L.)

The dicotyledonous family Cucurbitaceae and several others are unusual in having phloem on both sides of the xylem, within the vascular bundles of the stem. In most plants, only external phloem is present. In the larger vascular bundles of the herbaceous vine stem of pumpkin, the following zones can be distinguished: (i) primary external phloem (**1PE**); (ii) secondary external phloem (**2PE**); (iii) external cambial zone (**CE**); (iv) secondary xylem (**2X**); (v) primary xylem (**1X**); (vi) internal cambial zone (**CI**); and (vii) internal phloem (**PI**), both primary and secondary, with the primary phloem closer to the tip of the bundle. Sieve plates are just visible in some of the sieve-tube members.

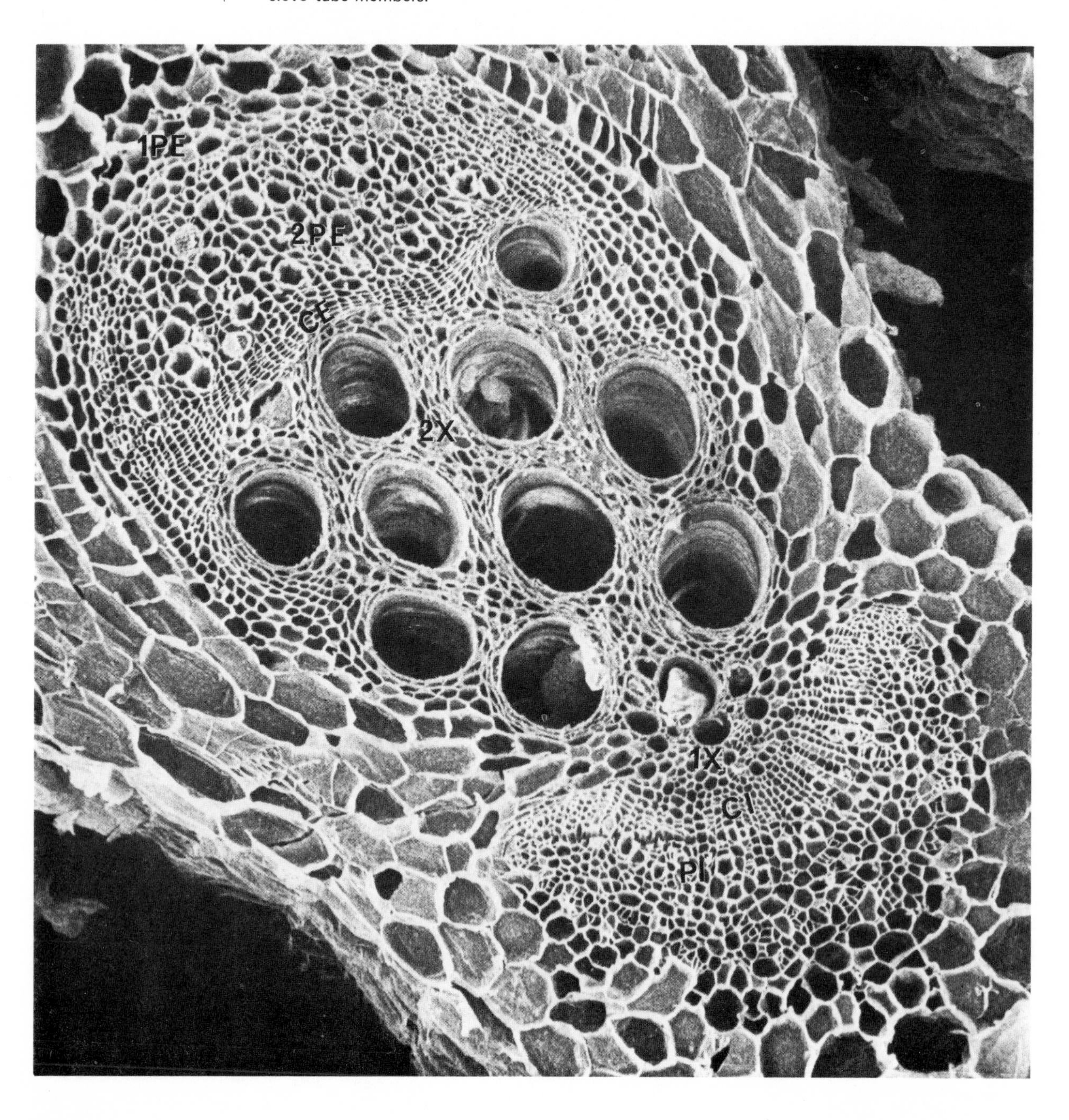

Transverse view of part of a corn stem *(Zea mays)*

Plate 113
X 100

The corn stem is typical of monocotyledonous plants in having numerous vascular bundles (**B**) scattered within a parenchyma ground tissue. The vascular bundles do not possess a cambium and therefore secondary vascular tissue is not formed. The primary phloem (**PH**) faces towards the outside of the stem. It consists of sieve tubes and companion cells. The primary xylem occupies the inner half of the bundle and consists of vessels and parenchyma.

M—large vessel in the metaxylem; **P**—large protoxylem vessel

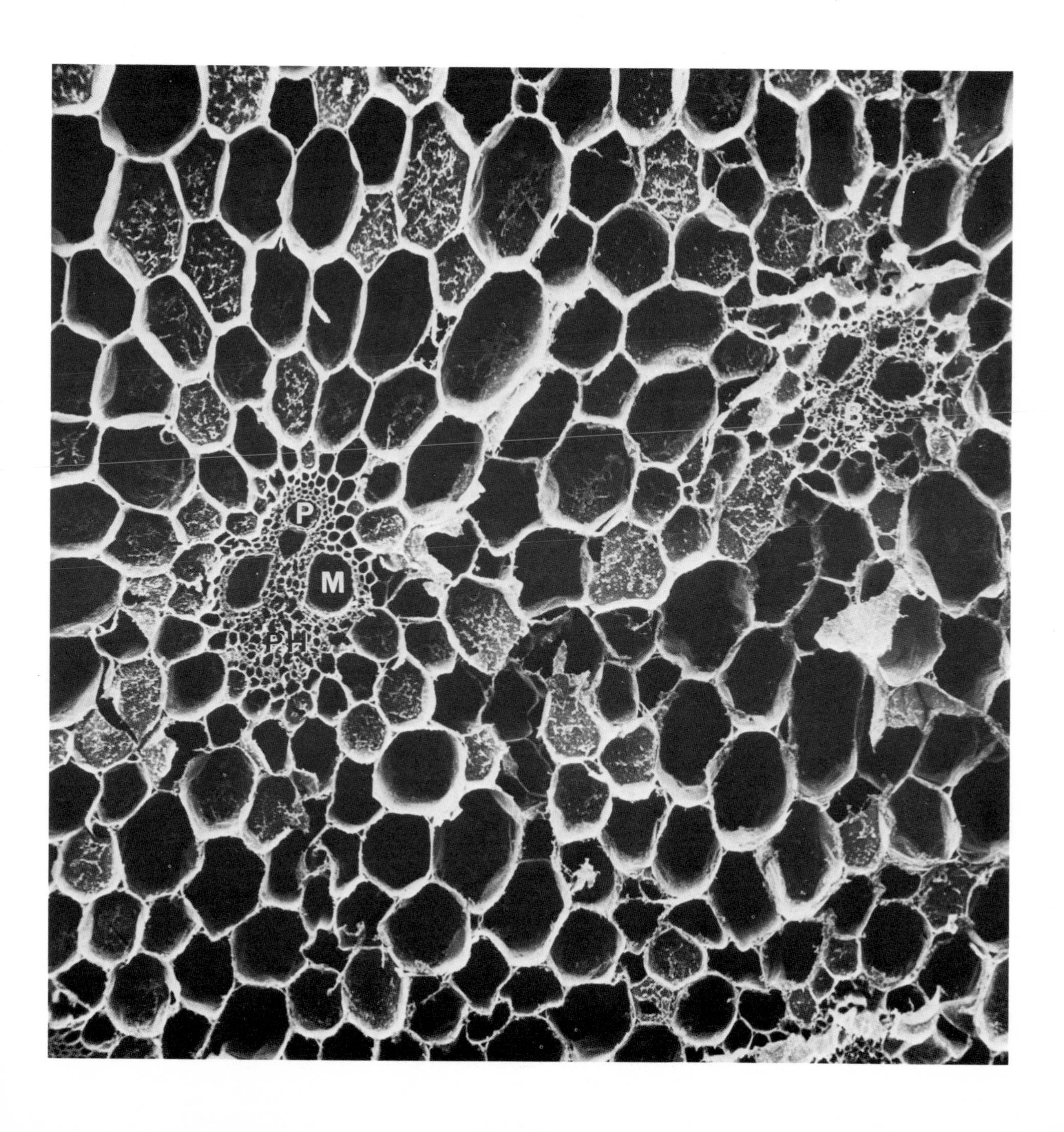

Plate 114

X 460

Cambial zone and bundle fibres in the stem of *Eupomatia*

In this view of a transverse section of *Eupomatia laurina* stem, through part of the cortex and vascular tissue, the cambial zone lies between the arrow heads. The cambium is meristematic tissue one cell wide, which is formed from the apical meristems of the shoot and root. It first becomes active within the vascular bundles and cuts off cells on its outer side, which mature into secondary phloem (**2P**). Derivatives on its inner side become secondary xylem (**2X**). In *Eupomatia* the cambium becomes active between the vascular bundles, which are close together, and secondary phloem and secondary xylem are formed there too. Although the cambium is only one cell in width, it cannot readily be distinguished from its immediate immature derivatives, and the term "cambial zone" is given to the cambium and these similarly shaped recent derivatives. At the stage illustrated, the cambial zone forms a cylinder in the stem (a ring as seen in transverse section) several cells thick.

A bundle of fibres (**F**) are at the outer face of the vascular bundle. Between the fibres and secondary phloem is the primary phloem (**1P**). Note the large parenchyma cells with small intercellular spaces in the cortex (**Co**).

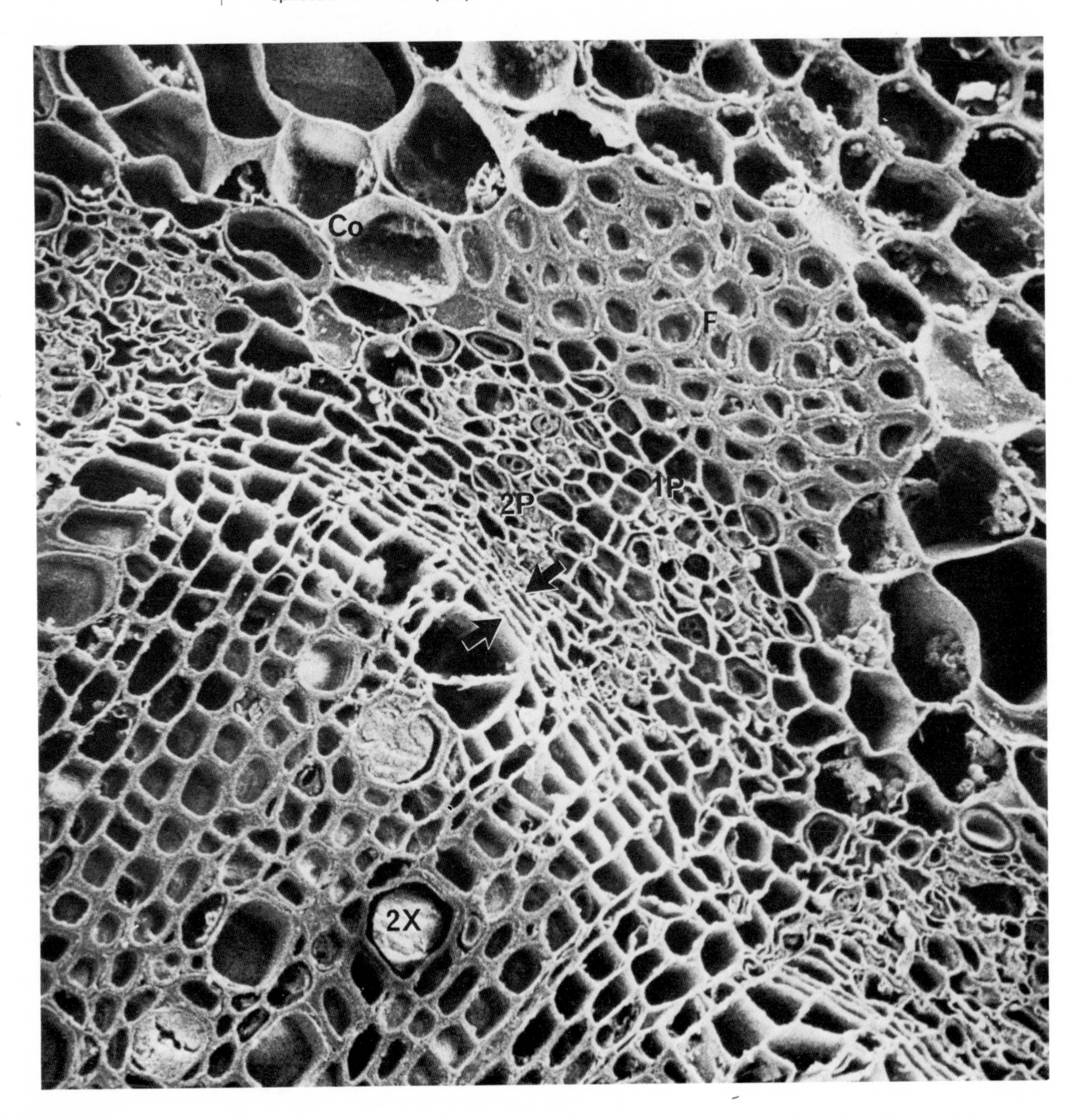

Ring porous wood of elm (*Ulmus* (Tourn.) L.)

Plate 115

X 160

Annual rings are very conspicuous in certain wood, in which large vessels are restricted to the early (spring) wood. In this transverse section of the stem of elm, the distance between the large vessels in the upper and lower parts of the photo represents a year's growth.

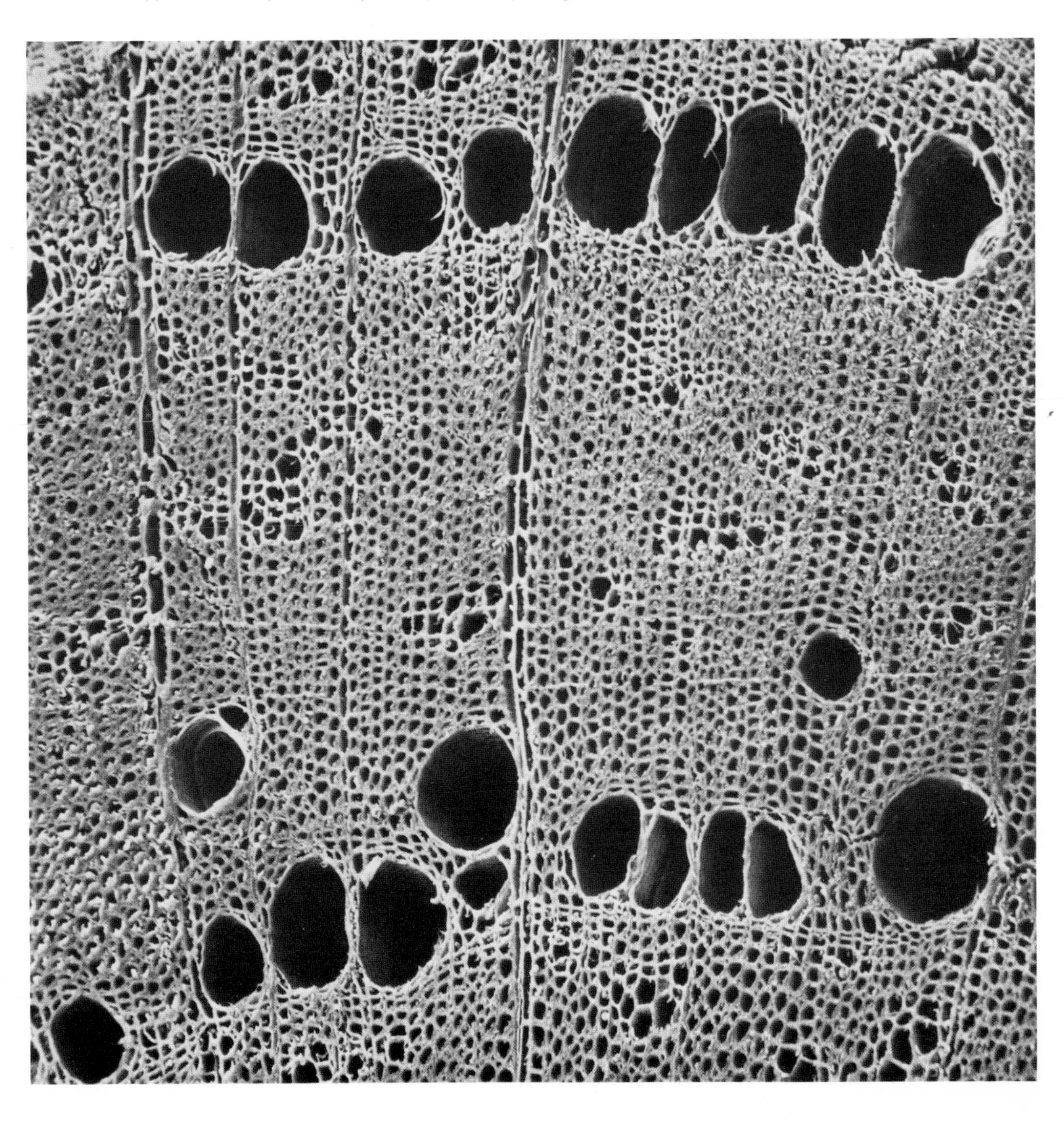

Plate 116
X 200

Diffuse porous wood of sycamore (*Acer pseudoplatanus* L.)

This transverse section of a sycamore stem illustrates the diffuse porous type of wood. Large vessels are scattered fairly uniformly throughout the wood (secondary xylem).

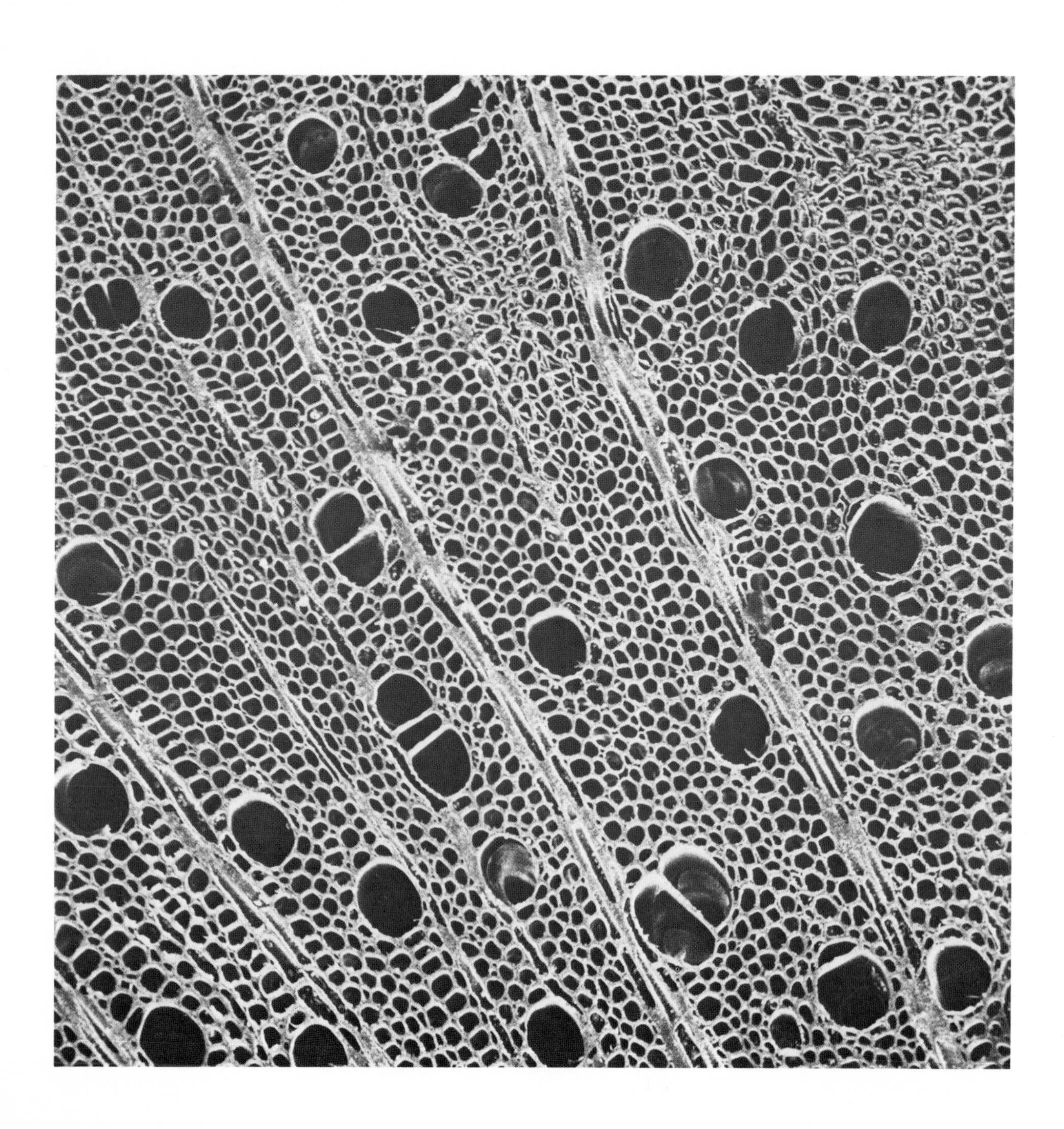

Root tip of radish seedling (*Raphanus sativus* L.)

Plate 117
X 170

At the end of the root there are root cap cells which protect the root tip as it grows through the soil. They become worn away during growth and are replenished by new root cap cells, which are outer derivatives of the apical meristem of the root. In contrast to the apical meristem of the shoot (Plate 111) the root meristem is in a subterminal position in the centre of the root tip. Derivatives from the sides and from the rear of the root meristem mature into the various primary tissues of the root, excluding the root cap.

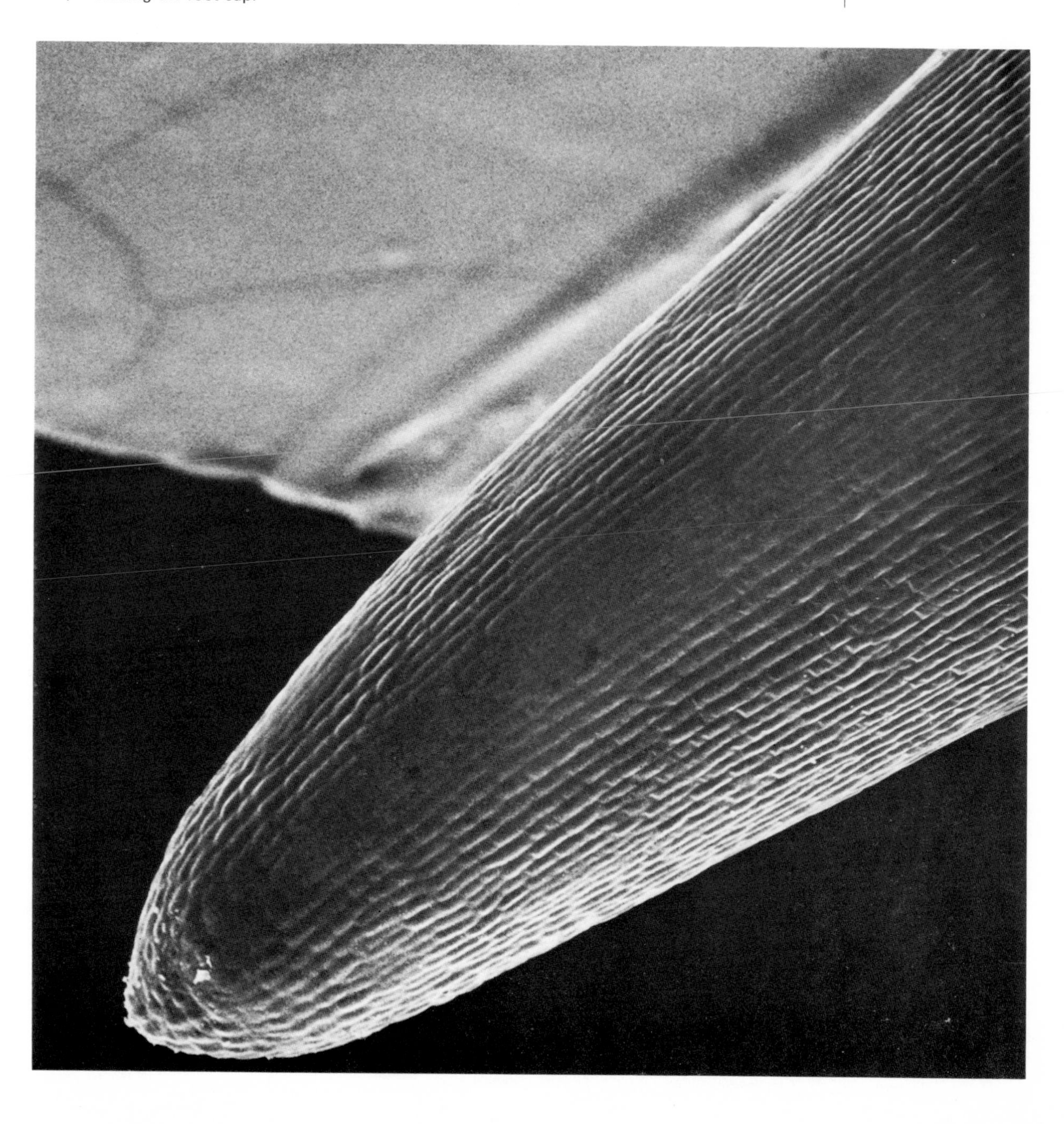

Plate 118
X 840

Root hairs of radish *(Raphanus sativus)*

Most young root surfaces are able to absorb water, but because the root hairs greatly extend the absorptive area of the root, most water absorption occurs through the root hairs. At an early stage in its development, a root hair cell is similar to other epidermal cells. It develops a small protuberance on part of its outer surface and this enlarges by growth at its tip into a tubular structure. Mature hairs are considerably longer than any in the photo. Root hairs generally remain alive for a few days only and new hairs are continually developing a millimetre or two from the tip of the root.

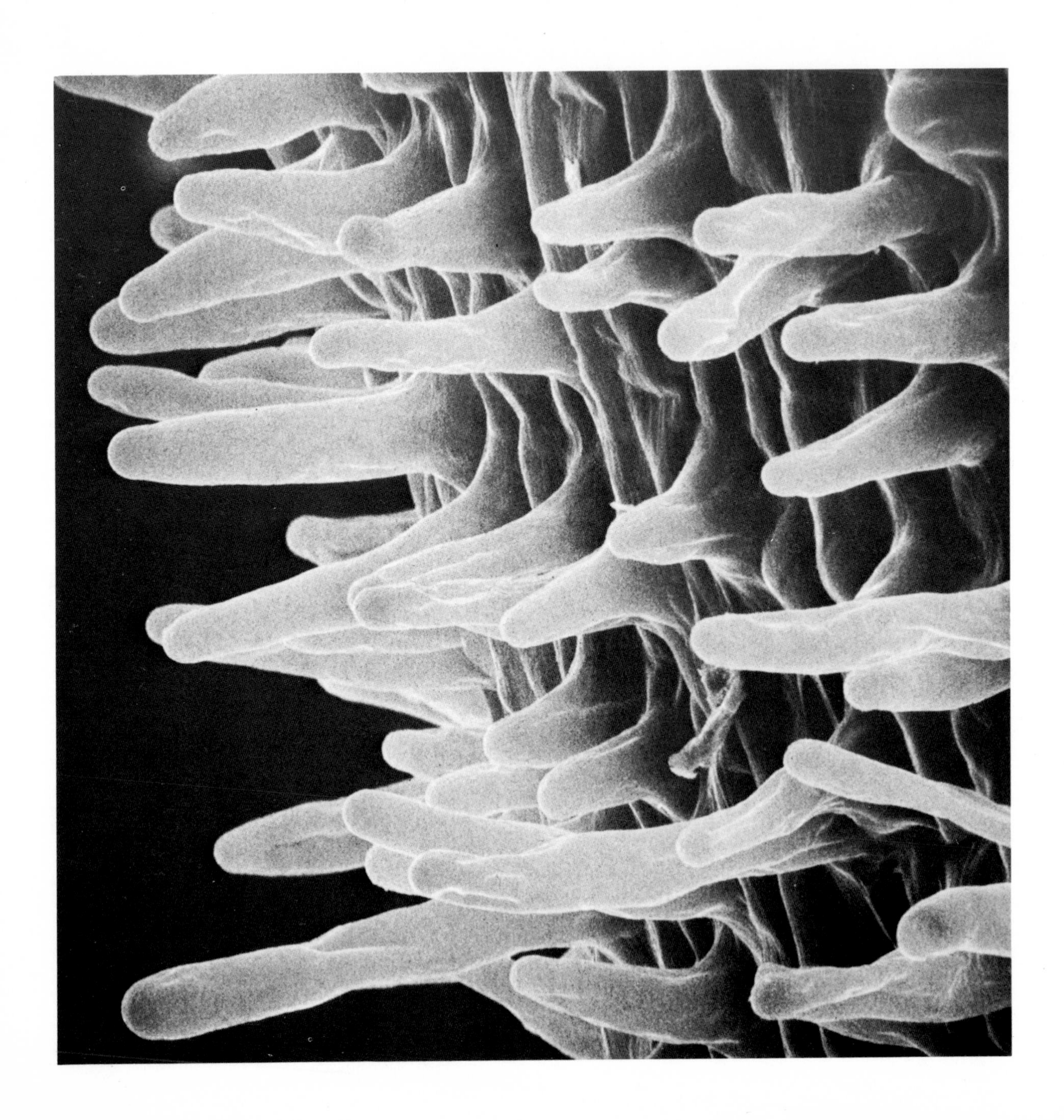

Transverse section of the central part of a corn root *(Zea mays)*

Plate 119
X 450

The vascular cylinder occupies the central region of this small root. The vascular tissue is surrounded by an endodermis (**E**), the cells of which have thick radial and inner walls but thin outer ones. The pericycle (**P**) is immediately beneath the endodermis. Vessels of the late metaxylem (**M**) are relatively large. The centre of the vascular cylinder is occupied by pith parenchyma cells (**C**). A typical feature of roots of monocotyledons is the large number of protoxylem groups outside the metaxylem. These protoxylem clusters alternate with the primary phloem. In this photo the primary phloem and protoxylem are insufficiently distinct from each other to enable them to be identified with certainty.

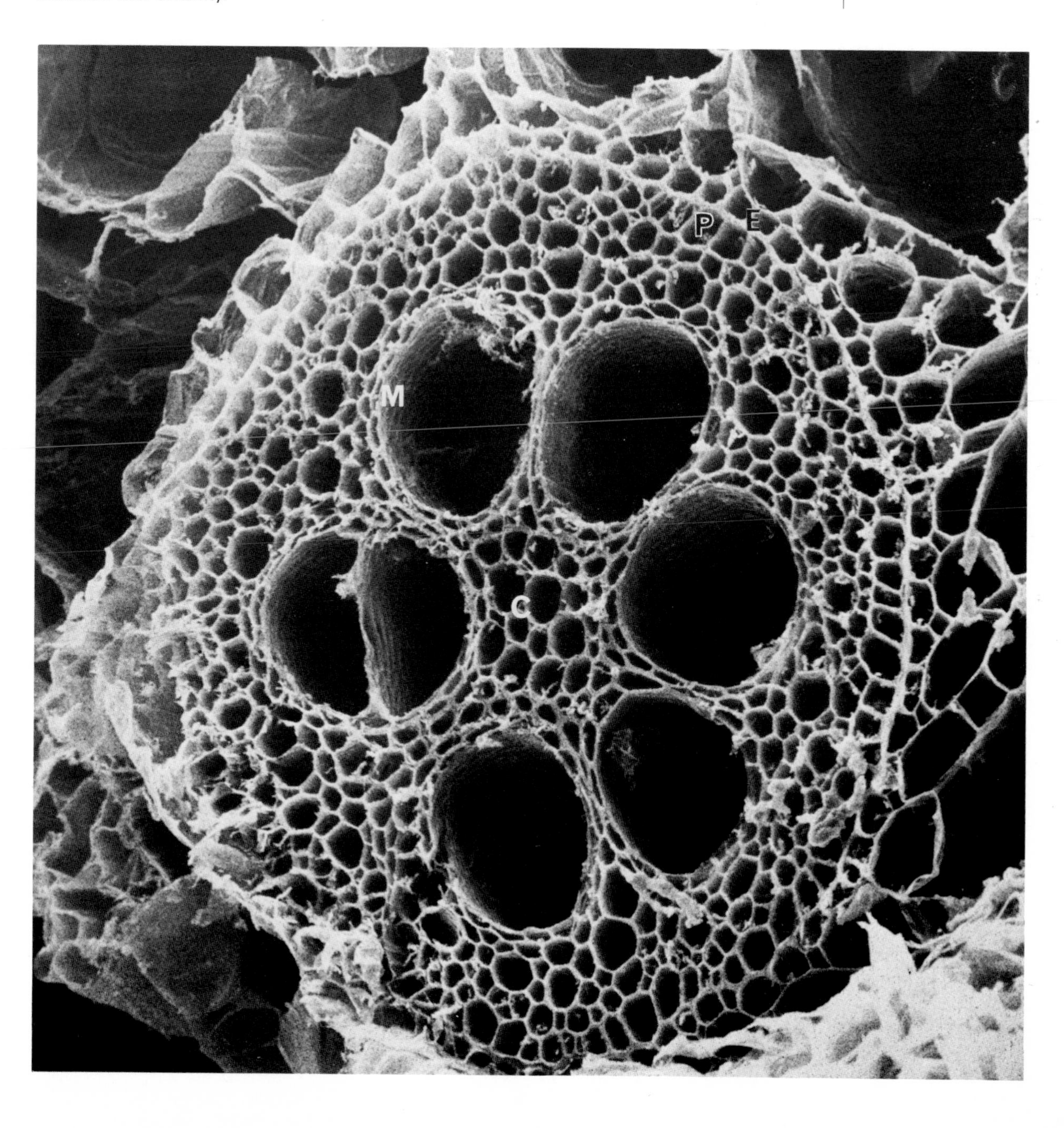

Plate 120
X 380

Leaf surface of rice (*Oryza sativa* L.)

Some grasses, for example, rice and oats (*Avena*), resemble *Equisetum* (Plate 62) and diatoms (Plates 5–8) in having silica deposits on the external walls of epidermal cells. These silica knobs occur just beneath a thin cuticle (Lewin and Reimann, 1969). The stomatal apparatus resembles that described for crabgrass (Plate 125). The boundary between the guard and subsidiary cells is not as sharply defined in rice, because of the platelet type of wax deposits on the surface of the cuticle. It can be seen that the guard cells lack the silica knobs. It is believed that silica helps to reduce unnecessary water loss and protects the plant against penetration by fungal hyphae. Experimental studies have shown that silicon-deficient rice shoots have a reduced fertility and retarded growth.

One of the stomata is open; the other almost fully closed. As is typical of monocotyledons, the stomata are parallel to each other.

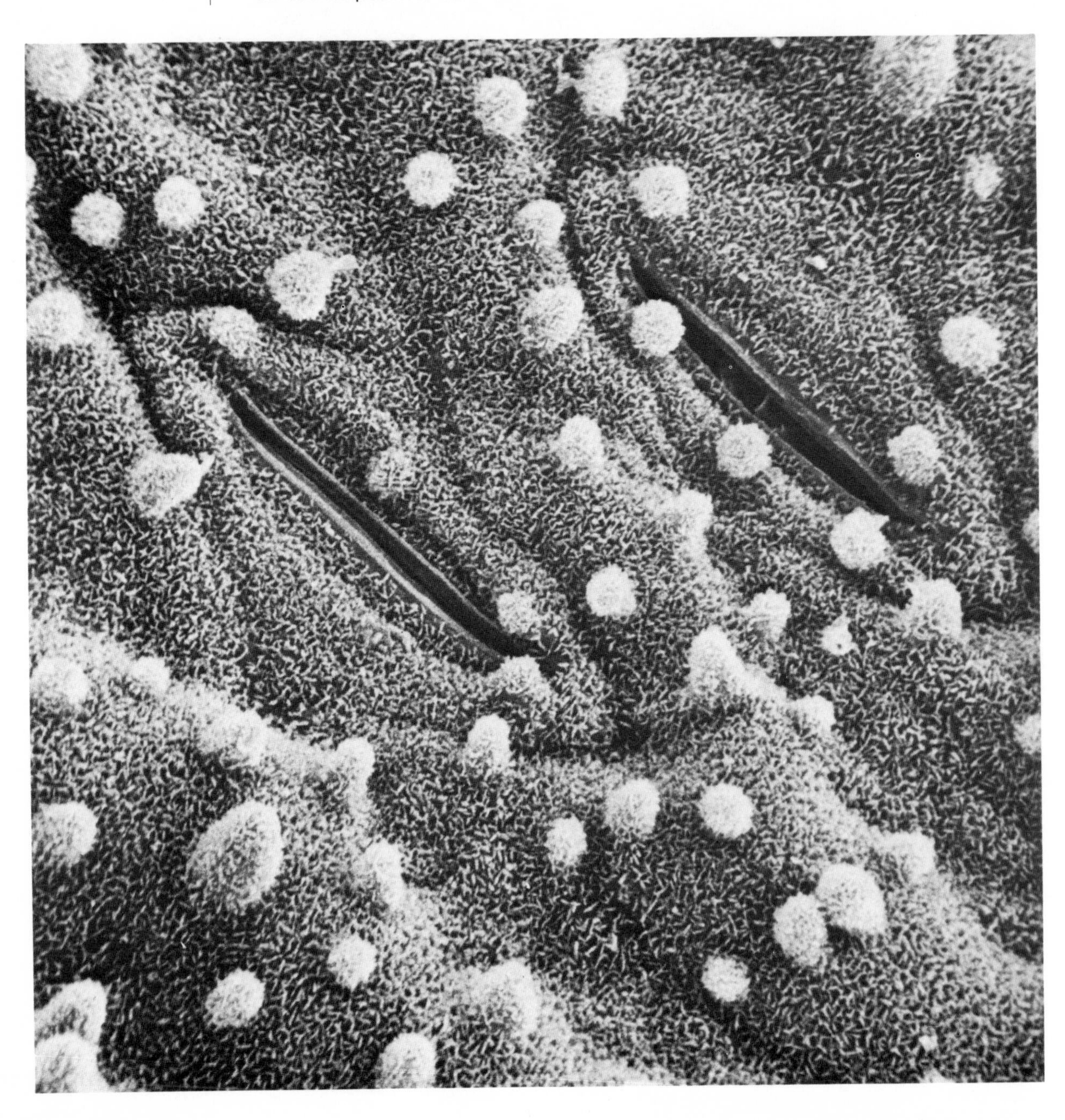

Leaf surface of cabbage (*Brassica oleracea* L.)

Plate 121
X 1,850

This is a view of the leaf surface of cabbage, showing three stomata. Above the surface of the cuticle are upright cylinders of wax.

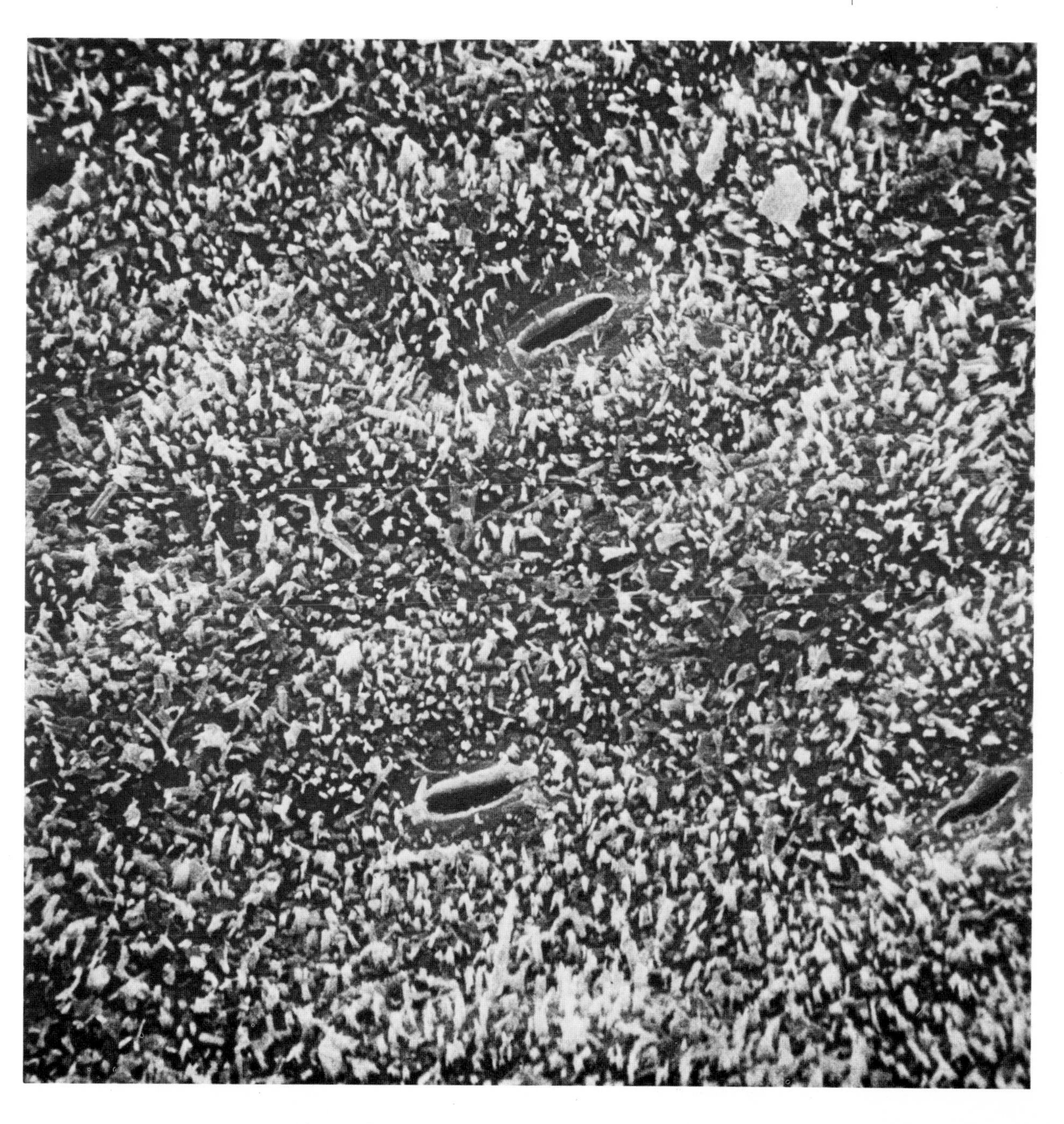

Plate 122

X 5,500

Wax on leaf surface of wheat (*Triticum* L.)

Rodlets of wax form a meshwork covering the surface of the leaf. Wax is synthesised within the cells of the epidermis and extruded through channels in the outer cell walls and cuticle and deposited on the leaf surface. A stoma, partly obscured by the wax rodlets, is at lower right.

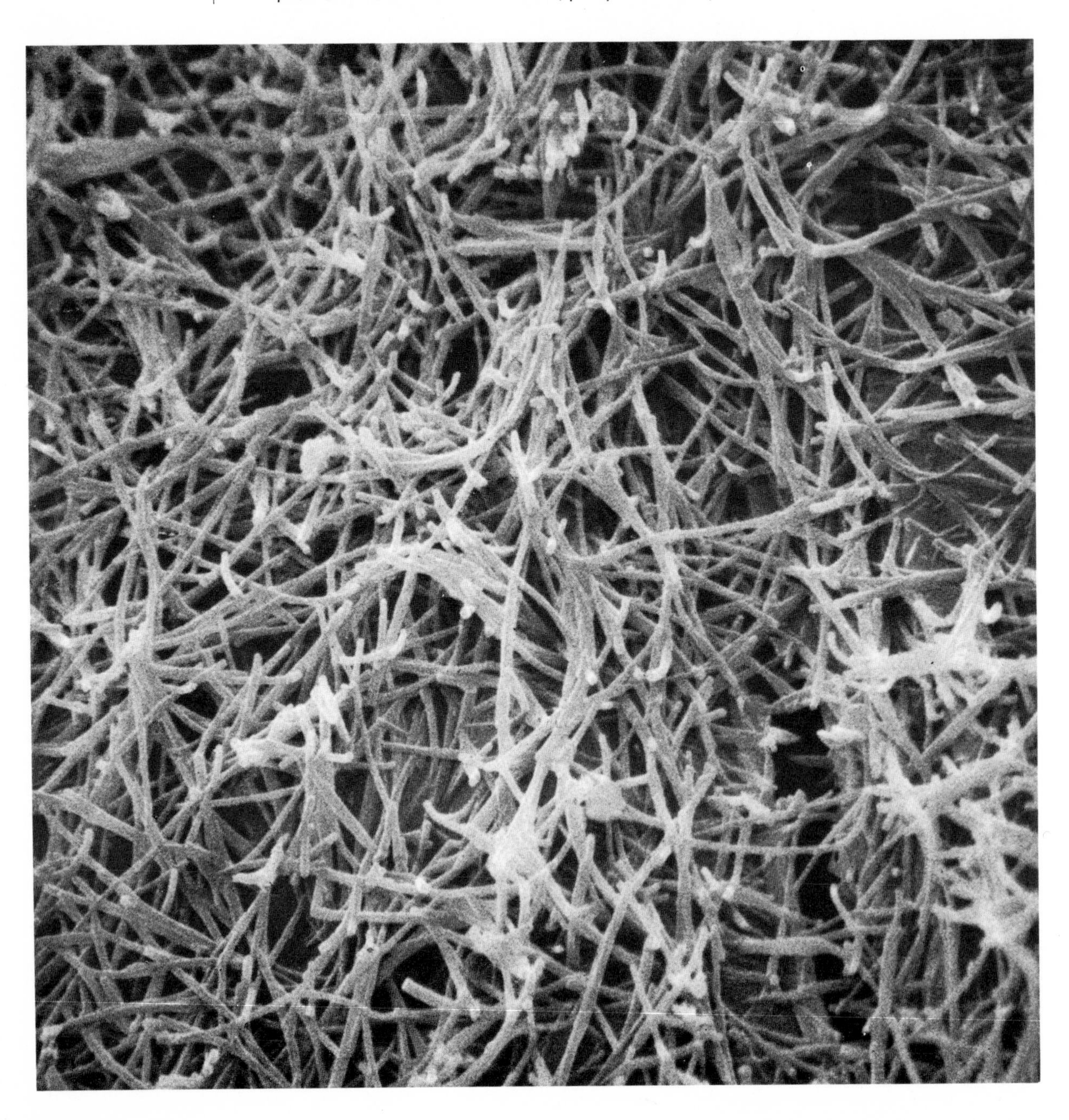

Wax on a banana leaf (*Musa* L.)

Wax on banana leaves is in the form of small curved ribbed sheets. The wax is synthesised in the epidermal cells, and extruded through the cuticle.

Plate 123
X 11,000

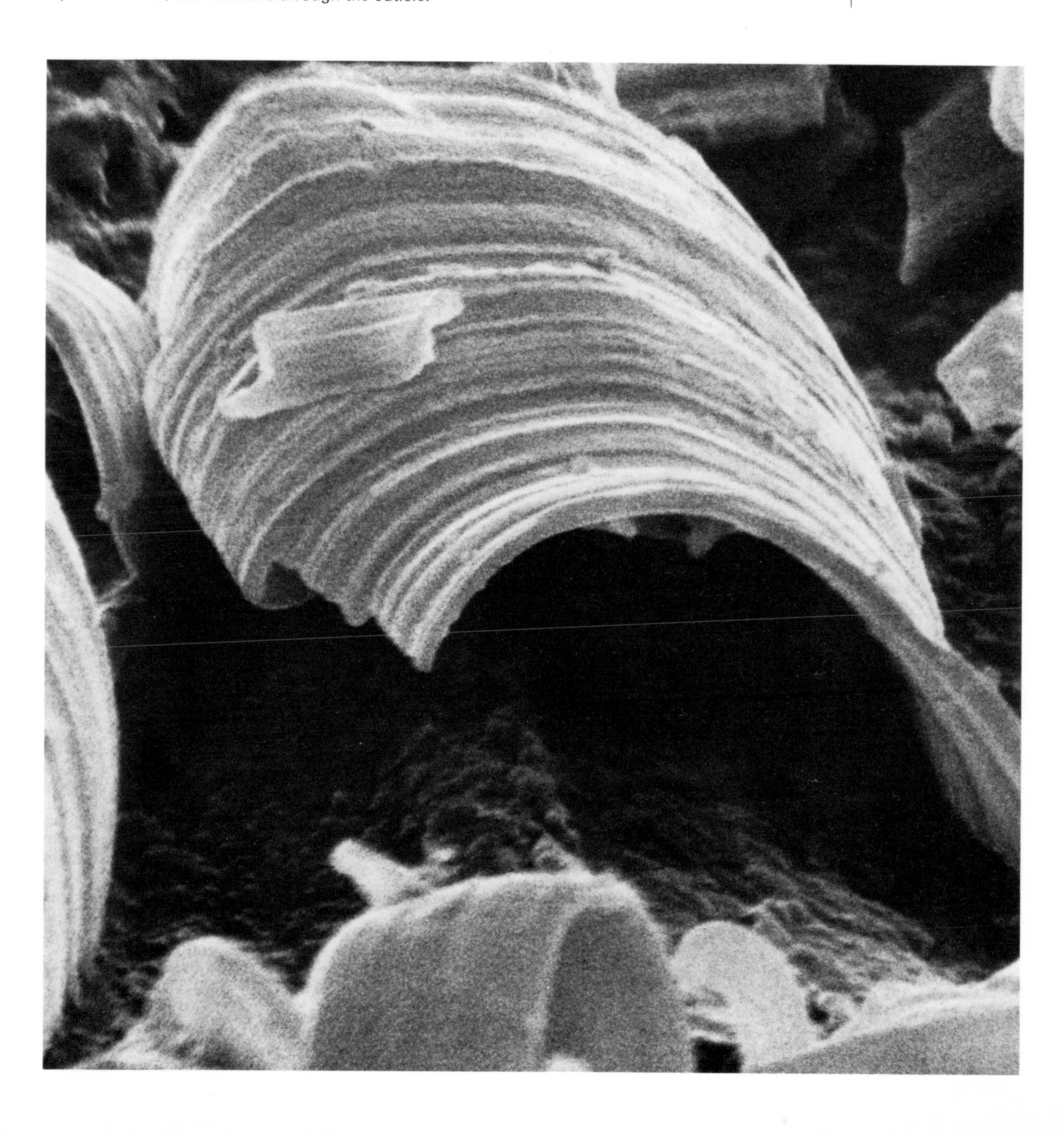

Plate 124
X 3,400

Guard cell of a stoma in lucerne (*Medicago sativa* L.)

In this transverse fracture of a leaf, a stoma was broken in half, revealing the shape of a guard cell (**G**) which has been broken at its ends. A small cuticular lip is apparent on the upper surface of the guard cell. Below the cell is a substomatal cavity, which is surrounded by mesophyll cells.

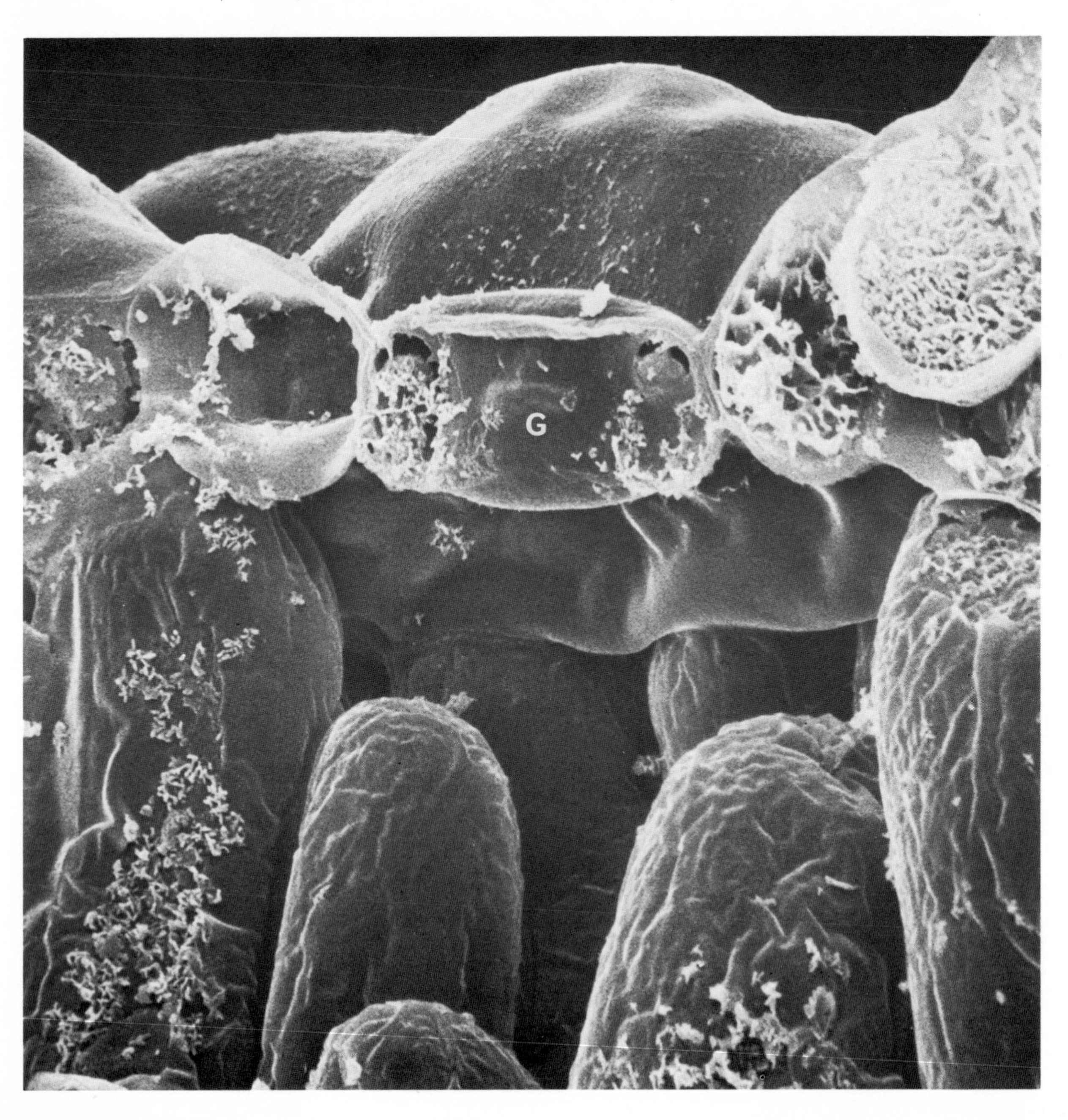

Stoma of crabgrass (*Digitaria sanguinalis* (L.) Scop.)

Plate 125
X 4,200

This illustrates the typical grass type of stomatal organisation. There are two subsidiary cells, one on each side of the guard cell pair. In contrast to the stomatal apparatus of *Equisetum* (Plate 62), for example, the four cells are not derived from the same parent cell. The stomatal apparatus is derived from three rows of epidermal cells, adjacent to one another. The guard cells are formed by division of a guard mother cell in the central row. (This cell is smaller than adjacent cells in this row.) Each subsidiary cell is formed by division into two unequal sized cells, of which the subsidiary cell is the smaller, in the adjacent row alongside the guard mother cell. Consult Esau (1965) for further details.

Plate 126 X 50

Stinging hairs on nettle (*Urtica incisa* Poir.)

Stinging hairs occur on the stem and leaves of this Australasian herb. The base of each hair is multicellular. Internal detail of a hair of *Urtica* is illustrated in Eames and MacDaniels (1947). Above the base, the limits of which are indicated by arrows in Plate 126, the hair comprises a single tapering thick-walled cell which has a globular tip. The bases of the hairs in the photo have been damaged. A close-up view of the tip is shown in Plate 127.

A sting from a single hair has the severity of a bee-sting and the region around the sting becomes numb. Stings from the New Zealand shrub nettle, *Urtica ferox,* have been severe enough to kill horses. This plant varies in its toxicity from one season and year to another. In December 1961, two men were severely poisoned by stings and one of them died. The poisonous principle is not definitely known (Thurston and Lersten, 1969). It causes paralysis of the central nervous system, leading in severe cases to blindness, unconsciousness and death.

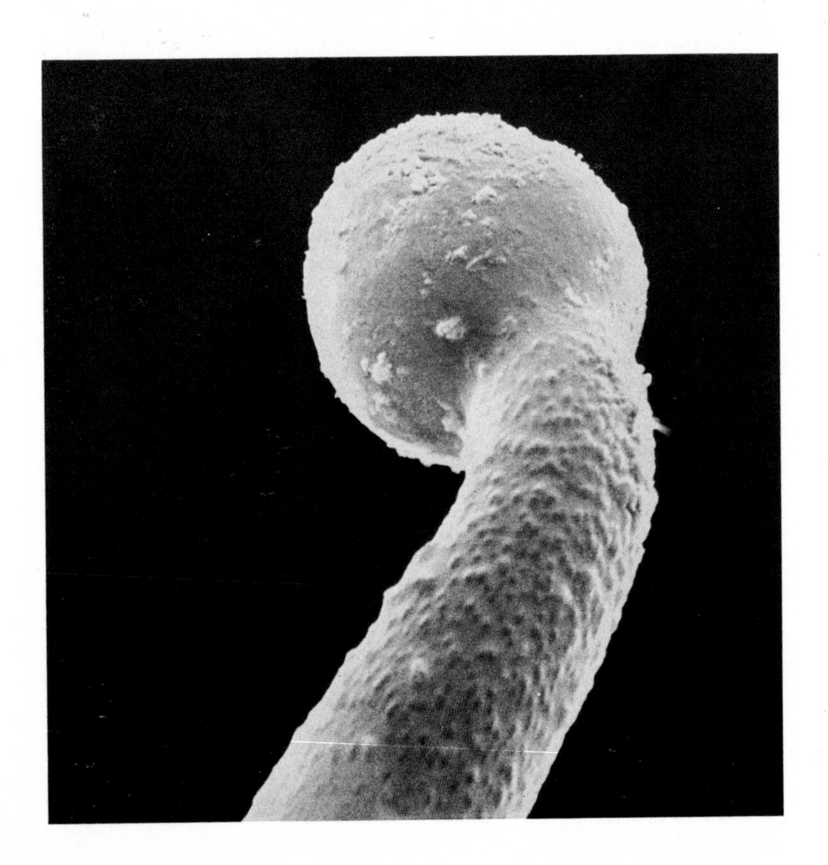

Plate 127 X 1,900

Peltate hair on the leaf of Norfolk Island hibiscus (*Lagunaria patersonii* G. Don.)

Plate 128
X 820

The leaves of this member of the mallow family (Malvaceae) are covered with multicellular peltate (shield-shaped) hairs. Each hair has a multicellular stalk, not visible in the photo, which is attached to the epidermis, and an expanded flattened head. The long narrow cells which form the peltate head are fused together for most of their length but are free at their tips. A secretory function has been attributed to these hairs (Metcalfe and Chalk, 1950).

Plate 129
X 440

Transverse view of a broad bean leaf (*Vicia faba* L.)

This illustrates a type of leaf typical of a mesophytic angiosperm, i.e. one which is adapted to growing under average conditions of water supply. Beneath the upper epidermis is a layer of vertically elongated palisade mesophyll cells (**P**). They are packed with chloroplasts and, due to some shrinkage during preparation of the material, the chloroplasts can be seen as small circular bulges within many cells. The lower half of the leaf contains the spongy mesophyll, a loosely arranged network of cells of irregular shape and large air spaces. There are also large air spaces between the palisade cells of broad bean and therefore much of the surfaces of internal leaf cells are exposed to the air. A small vein (**V**) is at centre left. Below the spongy mesophyll parenchyma is the lower epidermis, which can be seen in sectional and face view. A wider range of internal leaf anatomy as seen with the SEM is illustrated in Troughton and Donaldson (1972).

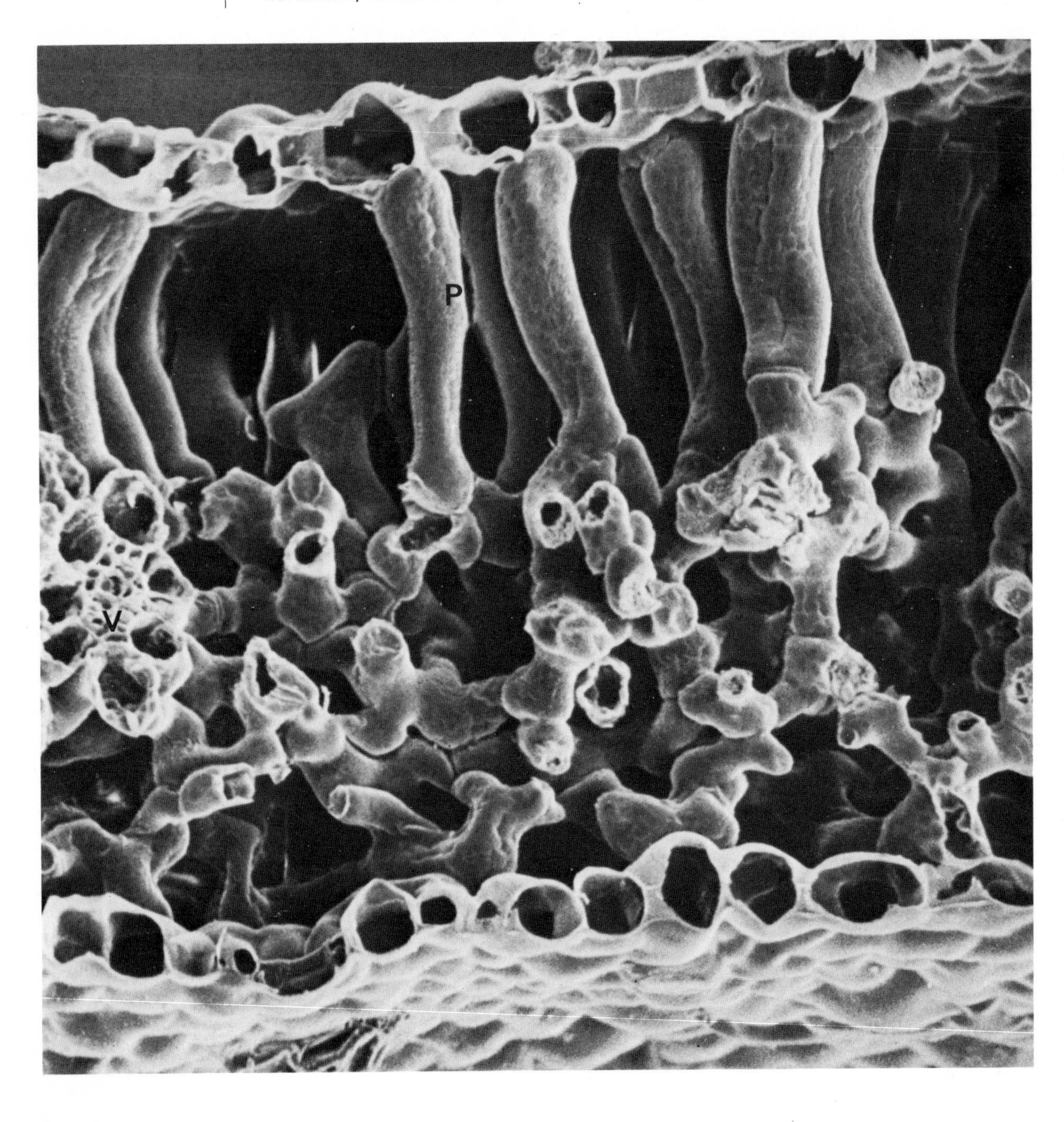

Transverse section of an oleander leaf (*Nerium oleander* L.)

Plate 130
X 440

This member of the family Apocynaceae illustrates the isobilateral type of leaf. In the contrasting dorsiventral type of leaf (Plate 129), the upper and lower parts of the leaf have a different cell structure, whereas the isobilateral type has a symmetrical internal structure. Palisade tissue (**P**) occurs above and below the spongy mesophyll (**S**). There is an exceptionally thick cuticle overlying the upper and lower epidermis in oleander. Beneath the epidermis is a hypodermis (**H**) several cells thick. The xylem (**X**) and phloem (**PH**) can be distinguished within the vein at centre.

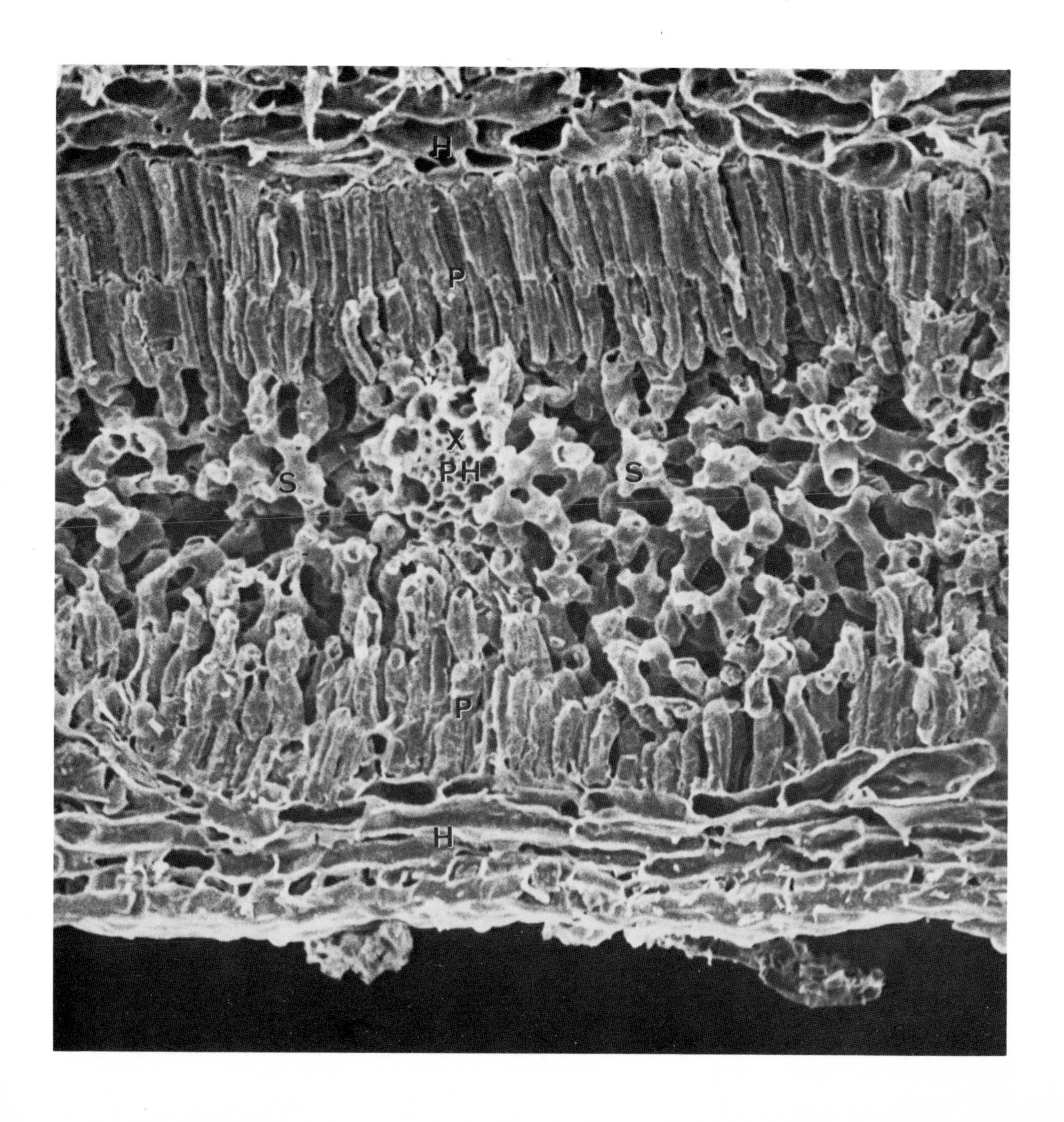

Plate 131
X 660

"Kranz" type of leaf anatomy in *Atriplex spongiosa*

Several anatomical features characterise "Kranz" type leaf anatomy. Two concentric layers of cells surround the vascular bundle, an outer mesophyll (**M**) and an inner parenchyma bundle-sheath layer (**B**). The inner cell layer contains chloroplasts and often has thickened cell walls.

"Kranz" anatomy occurs in several families of monocotyledons and dicotyledons. In the monocotyledons there is the following progression or apparent evolutionary series in leaf structure. The first type lacks a parenchyma bundle-sheath and has irregular, loosely arranged mesophyll cells (festucoid line). There is an advance towards a compact mesophyll, with elongated closely arranged chlorenchyma cells radiating out from the vascular bundle (bambusoid and arundinoid lines). Further specialisation of the chlorenchyma tissue gave the "Kranz" type anatomy, which is most highly developed in the grasses, sub-family Eragrostoideae.

"Kranz" leaf anatomy is a characteristic of plants with the recently established pathway of photosynthesis, known as the C_4 pathway (Hatch and Slack, 1970). Plants without "Krantz" anatomy have the Calvin (or C_3) pathway of photosynthesis, or Crassulacean Acid Metabolism.

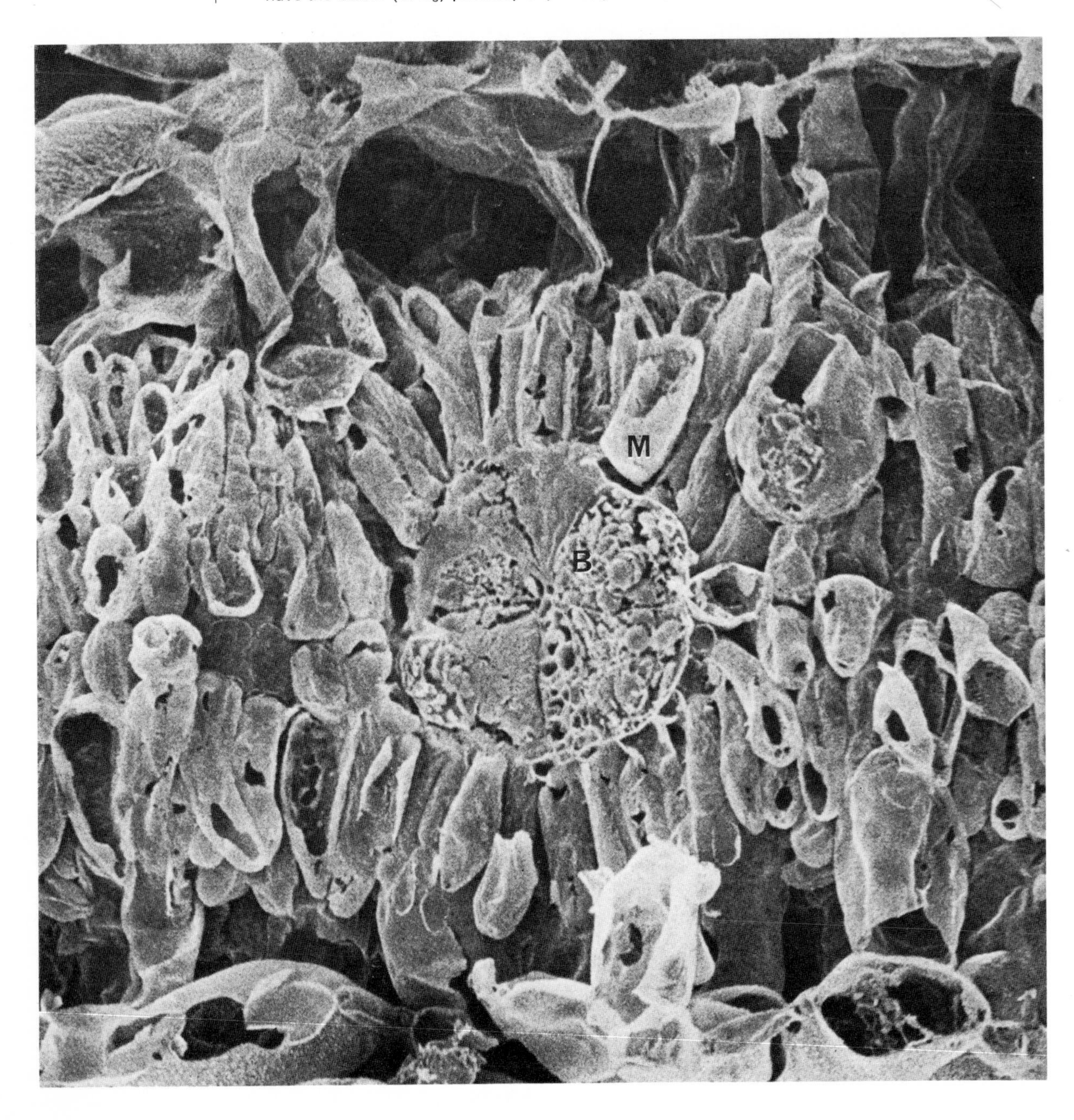

Longitudinal fracture of a crabgrass leaf *(Digitaria sanguinalis)*

Plate 132

X 1,200

Digitaria sanguinalis has "Kranz" leaf anatomy and C_4 photosynthesis (see Plate 131). The leaf was fractured in liquid nitrogen and most mesophyll cells were removed from the bundle-sheath cells (**B**). Some mesophyll cells remain (**M**). The outer surface of the bundle-sheath cell is shown in this photo.

The interface (**I**) between the mesophyll and bundle-sheath cells is visible. Removal of the mesophyll cells was not complete and cell remains define the area of the interface. A substantial area of the bundle-sheath surface is not occupied by cell interface and is therefore exposed to intercellular air.

The bundle-sheath cells form a cylinder surrounding the vascular bundle. The passage of carbon compounds into sieve tubes from the mesophyll cells, and water and nutrients from the xylem to the epidermis, must occur through the mesophyll bundle-sheath cell interface. In some plants, especially grasses, there is a suberised layer in the cell wall which seals the bundle-sheath cell. Exchange across the interface is primarily through the plasmodesmata.

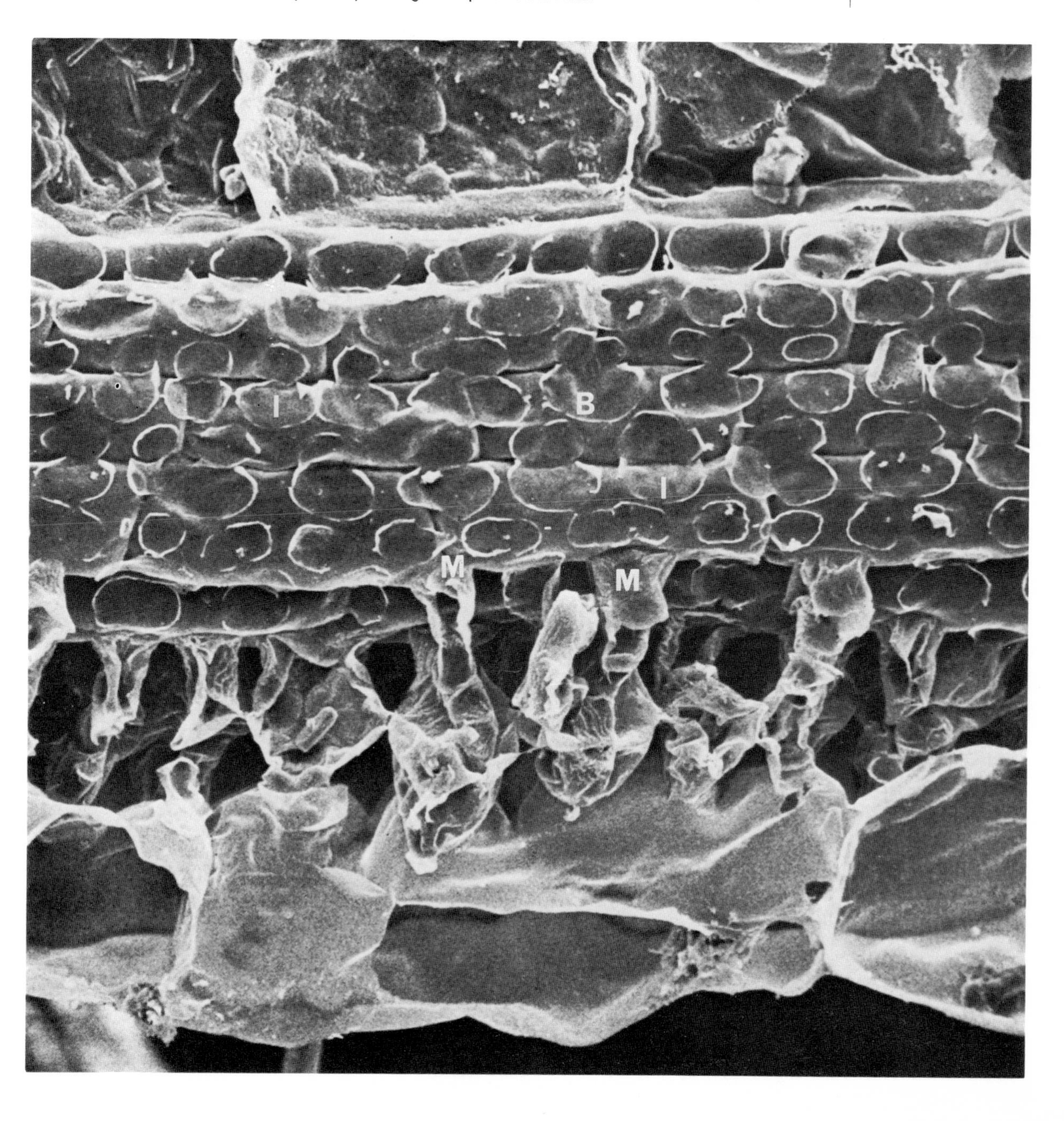

Plate 133

X 26,000

The primary pit-field and plasmodesmata in the bundle-sheath cell of crabgrass *(Digitaria sanguinalis)*

Plasmodesmata connect the cytoplasm of adjacent cells of higher plants. Aggregation of as many as 30 plasmodesmata occurs in localised depressed regions of the primary cell wall called primary pit-fields.

This view of the outer surface of a bundle-sheath cell illustrates a primary pit-field at the mesophyll—bundle-sheath cell interface of a leaf of crabgrass. Plasmodesmata do not occur on the bundle-sheath cell surface outside the interface, which is defined by the frill of cell wall material left when the mesophyll cell was removed.

Structural details of plasmodesmata are beyond the resolution of the SEM. Other microscopic techniques show that the pore traversing the cell wall is 30 to 100 nanometres in diameter. The pore is lined by the plasmalemma (cell membrane) and contains other membrane material, probably derived from the endoplasmic reticulum.

There is some similarity in form between plasmodesmata and sieve pores in the phloem. The sieve pores in *Cucurbita* (Plate 108) develop from regions where there are plasmodesmata (Esau, 1969).

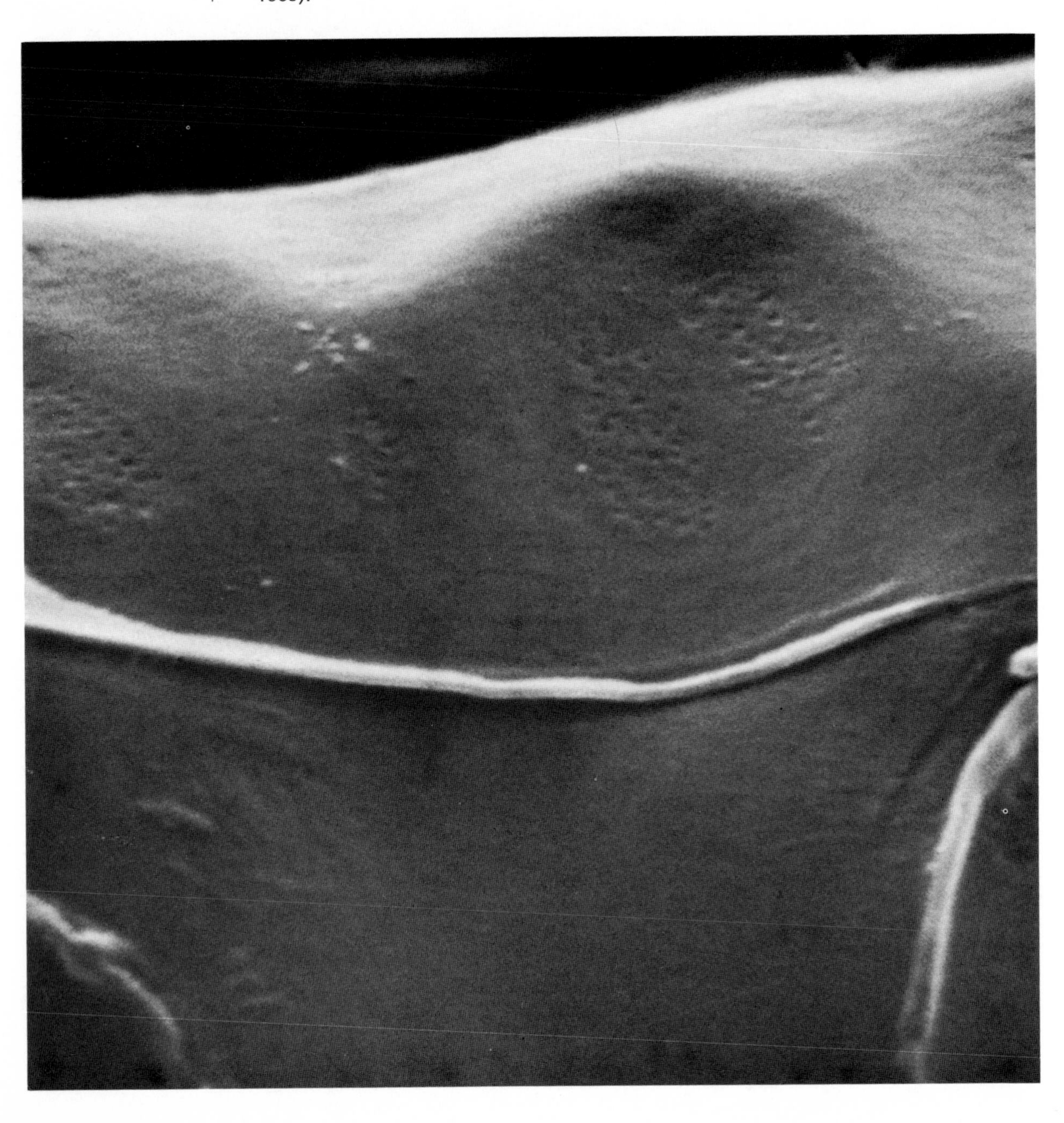

Inflorescence development in *Coleus*

Plate 134
X 300

At the top of the photo is the dome-shaped primary inflorescence meristem. Near its base a pair of oppositely arranged leafy bracts have been initiated. As with the leaves of *Coleus*, each pair of bracts is at right angles to the previous pair. A secondary inflorescence meristem (**S**) arises in the axil of each bract. Several flowers are formed from each of these inflorescence meristems. The developing inflorescence can be readily distinguished from the vegetative bud (Plate 111) because the bracts grow comparatively slowly, and even the tips of the fourth youngest pair in the photo have not yet reached the level of the primary inflorescence apex. Each inflorescence arises in the axil of a leaf.

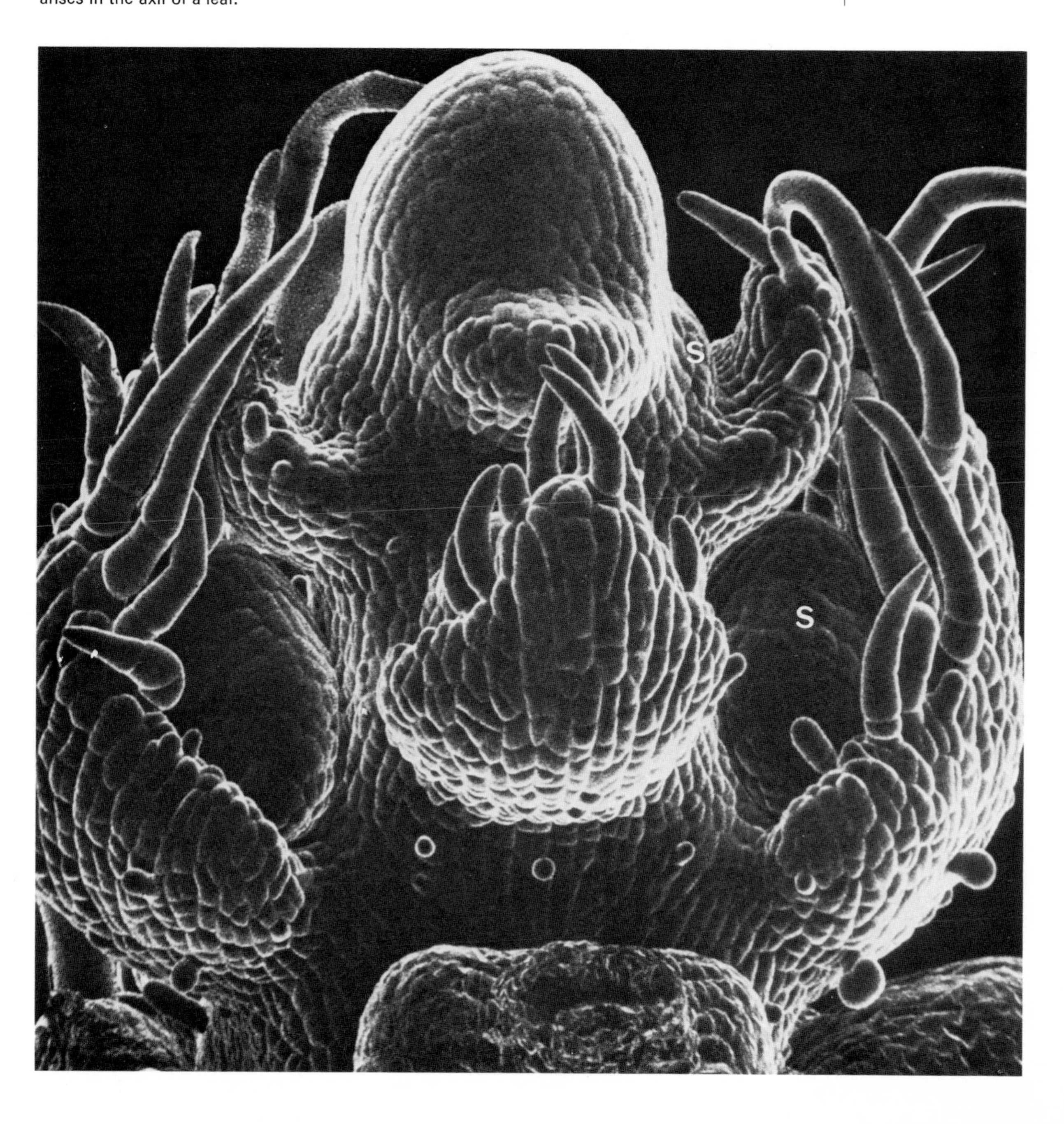

Plate 135 X 210

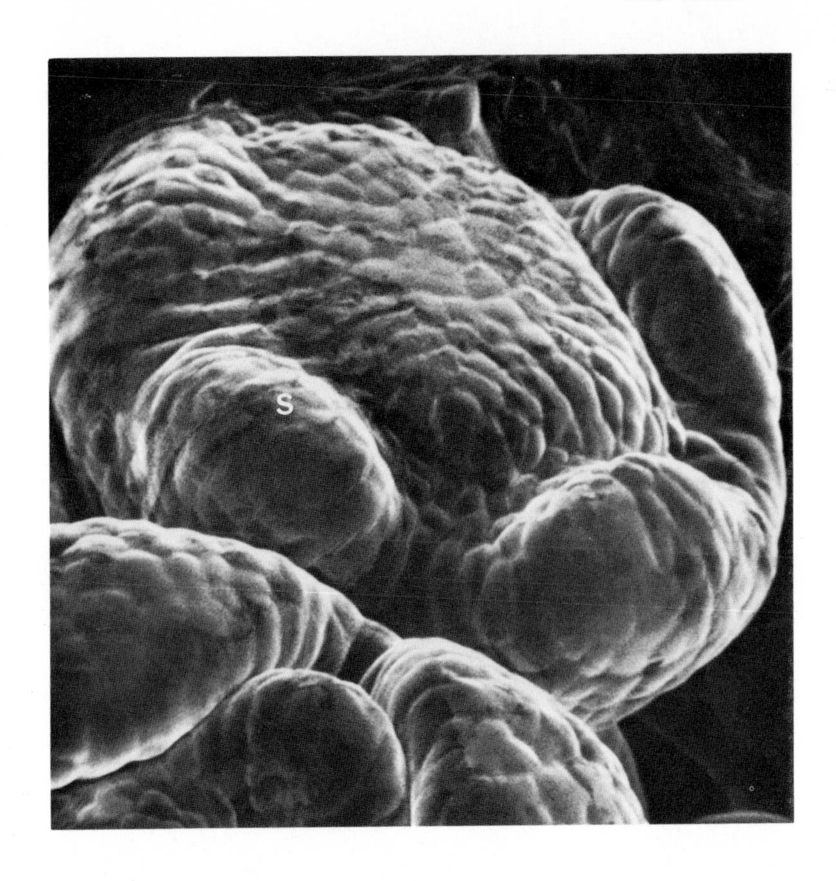

Floral development in *Coleus*

The floral apical meristem is also dome-shaped in *Coleus*. A ring of five sepal primordia are first initiated from the base of the floral apex (Plate 135). Part of an older flower bud is at lower left. The floral apex then increases in size and a whorl of five petal primorida develop internal to and alternating with the sepals (Plate 136). A whorl of four stamens then arise from the flanks of the apex, inside the petals and alternating with them. In flowers of this particular dicotyledonous family (Labiatae) the fifth stamen is usually absent. The space where the "missing" stamen would be is at what is morphologically the top of the flower—next to the primary inflorescence axis. Finally the gynoecium is formed in the centre of the flower (Plate 137). It consists of two fused carpels in this family. In the early growth of the gynoecium the small residual floral apex is "used up."

S—sepal; **P**—petal; **G**—gynoecium

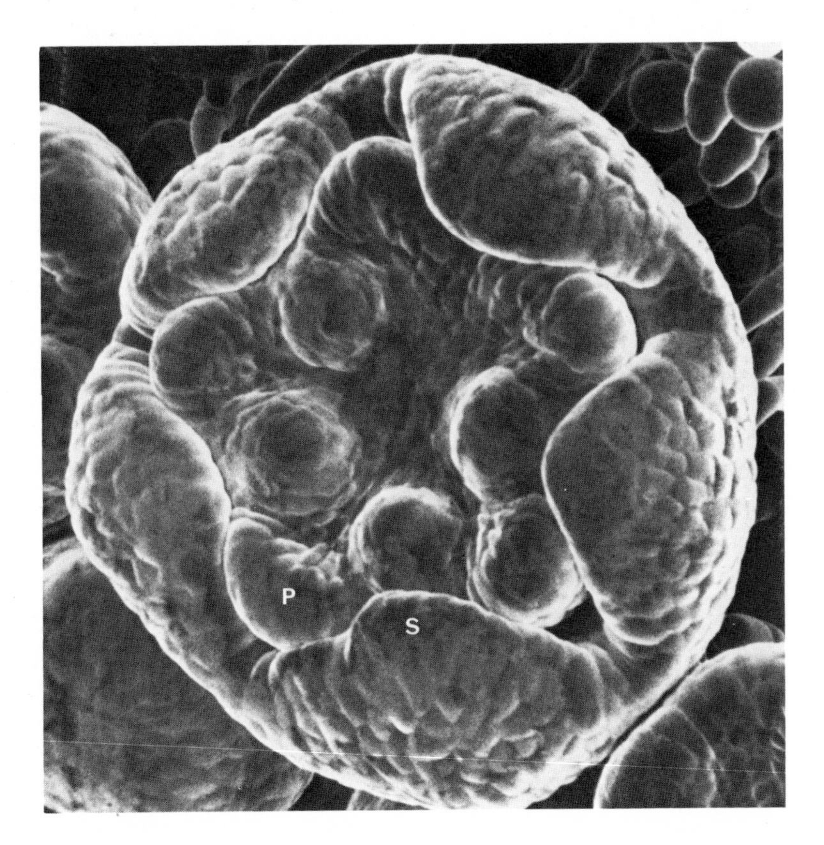

Plate 136 X 320

Plate 137

X 440

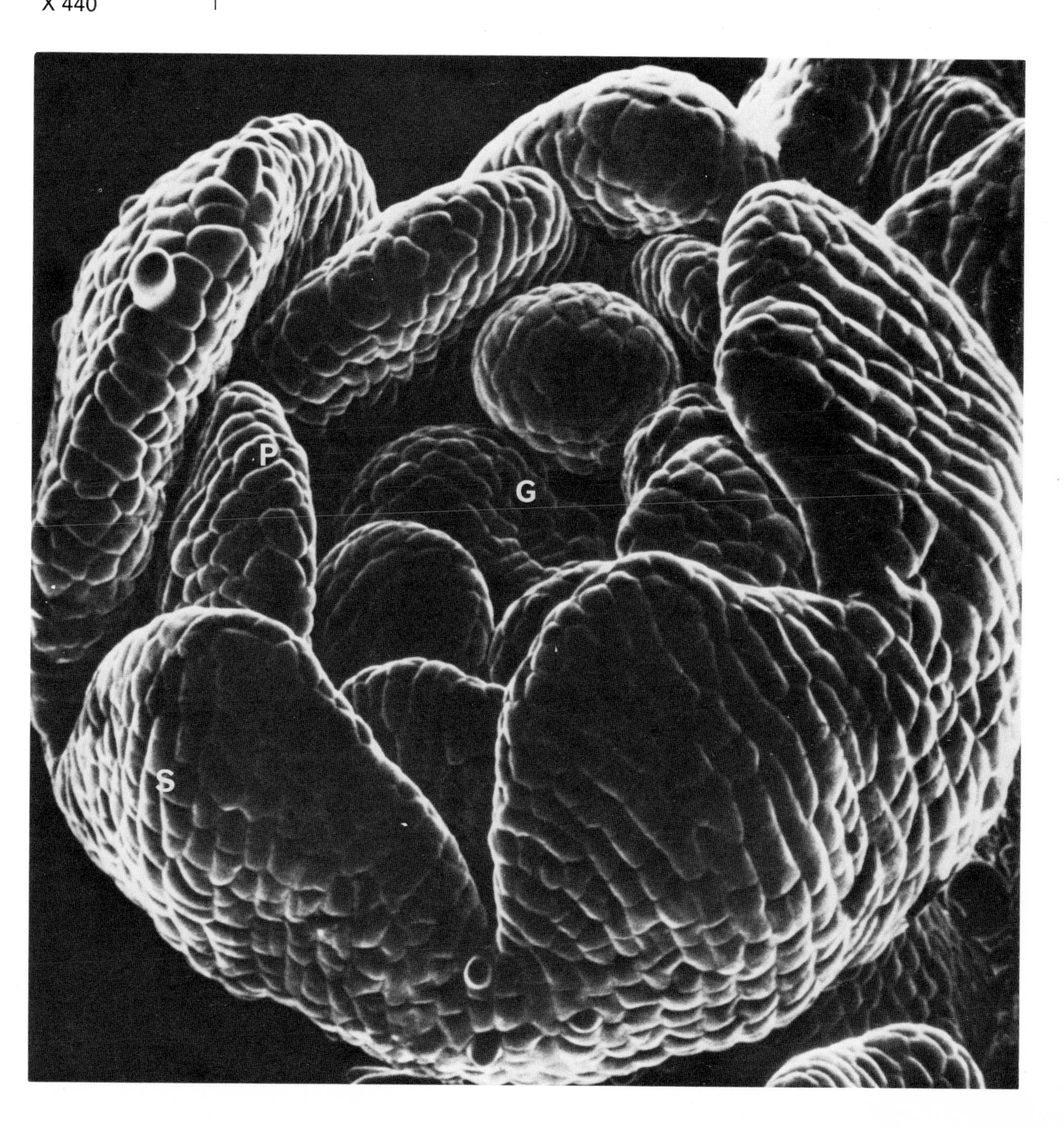

Plate 138
X 50

The flower of sea alyssum (*Lobularia maritima* (L.) Desv.)

There is little variation in the number and arrangement of floral parts in the mustard family (Cruciferae), which includes alyssum. In most members of the family there are four sepals (**K**), and four petals (**P**), which lie outside six stamens (**A**). The stamens in the photo have shed their pollen. In the centre of the flower is a gynoecium consisting of a stigma (**S**), a short style and an ovary (**O**). It is a compound structure which has been derived from two carpels. Two of the stamens, at the top and lower parts of the photo, lie outside the other four.

Petal surface of *Diosma alba* Thunb.

Plate 139
X 1,800

The protuberant surfaces of the epidermal cells of this petal form an attractive pattern. The epidermal cells are covered by a cuticle and the ridges are probably wax deposits. *Diosma* is a member of the citrus family (Rutaceae).

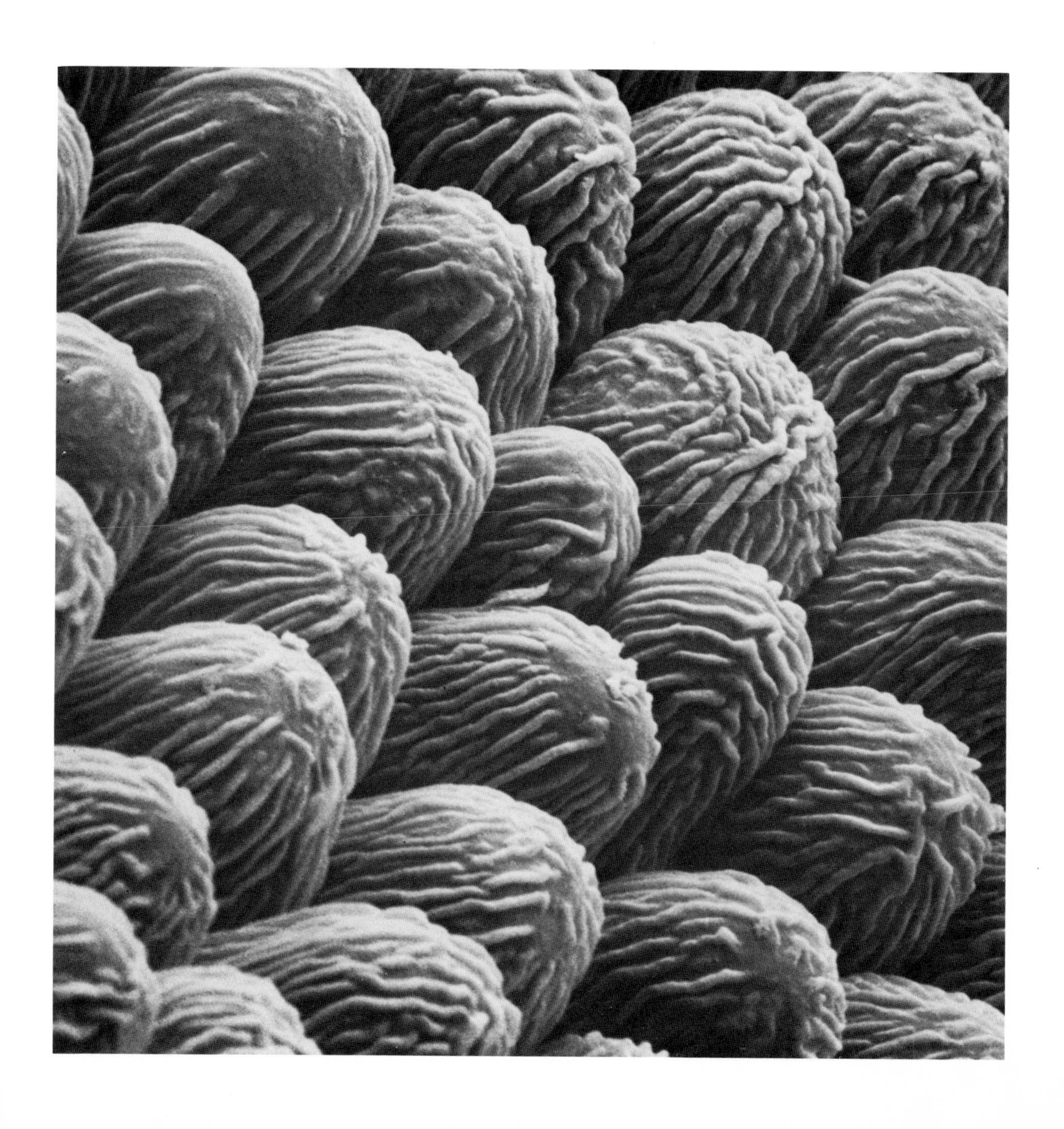

Plate 140
X 300

Stigma of primrose (*Primula* L.)

There is a spherical stigma (**S**) at the tip of a long style (**SY**), in the centre of the primrose flower. The base of the style is attached to the top of a compound ovary which has been derived from five carpels.

The epidermal cells of the stigma are protuberant and these projections help to retain pollen grains which fall on the stigma.

Most species of *Primula* have two types of flowers, which are always on different plants. One type, the so-called "pin" flower, has a long style with the tips of the stamens below the level of the stigma. The other "thrum" type has a shorter style and the pollen sacs of the stamens lie above the stigma. The two types occur in approximately equal numbers. It has been suggested that this arrangement of short-styled and long-styled flowers favours cross-pollination by insects, in that transfer of pollen from a pin to a thrum flower and vice versa is more easily accomplished than transfer of pollen from a pin (or thrum) flower to another pin (or thrum) flower. Such cross-pollination produces genetic variability of the offspring and this provides greater evolutionary potential. It has been found that pin and thrum flowers are partially self-incompatible. Thus, more viable seeds are formed when one form is fertilised with pollen from the other form.

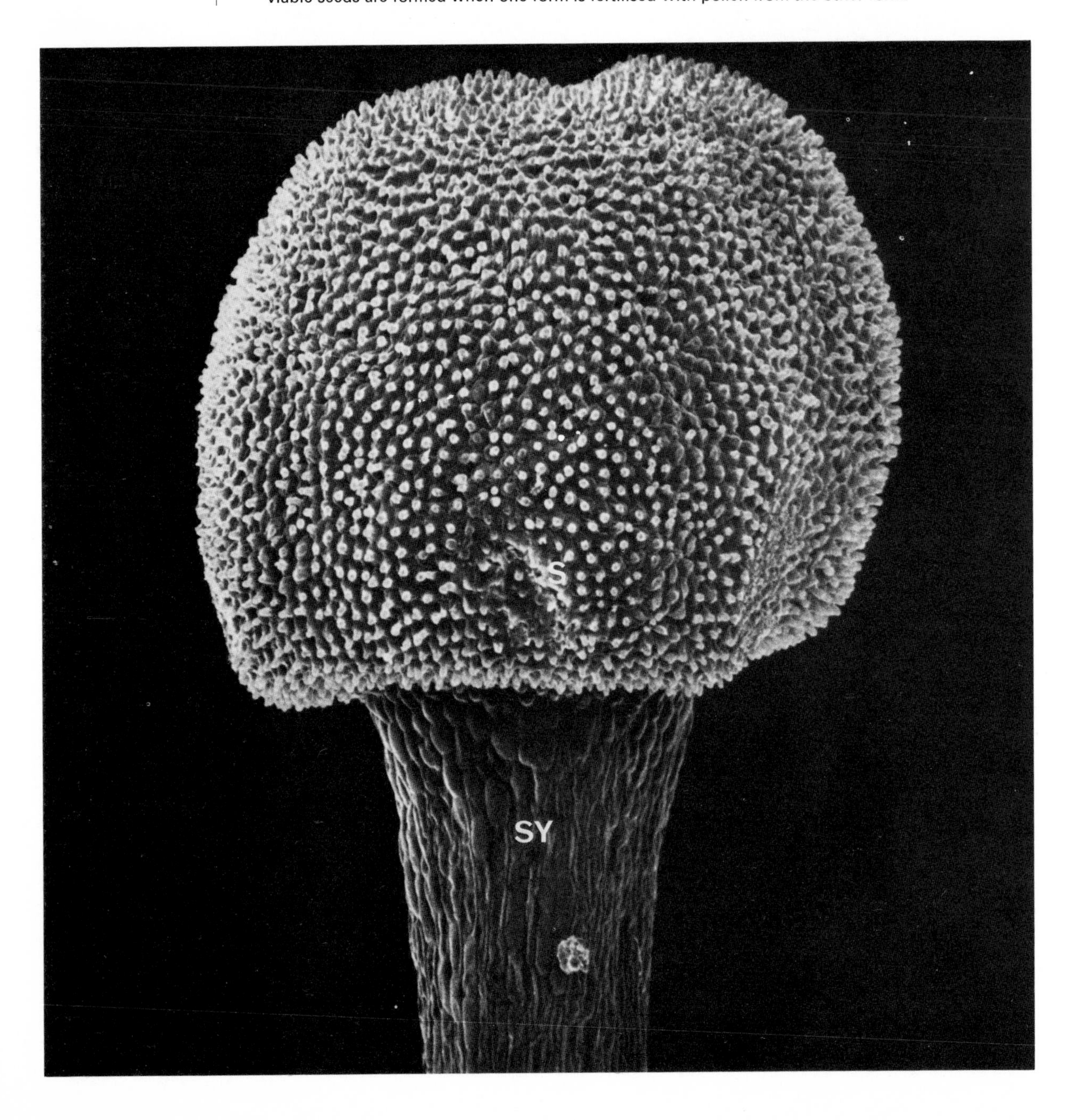

Thrum pollen on a pin stigma in primrose (*Primula*)

Plate 141
X 2,000

In addition to the differences between pin and thrum flowers already mentioned (Plate 140), the pin (long-styled) flowers have longer epidermal papillae (**P**) than thrum (short-styled) flowers. The pollen from short-styled flowers, shown above, is more spherical and larger than that from pin flowers (Plate 142). Some of the grains have germinated and a pollen tube (**T**) extends from each of these into stigmatic tissue.

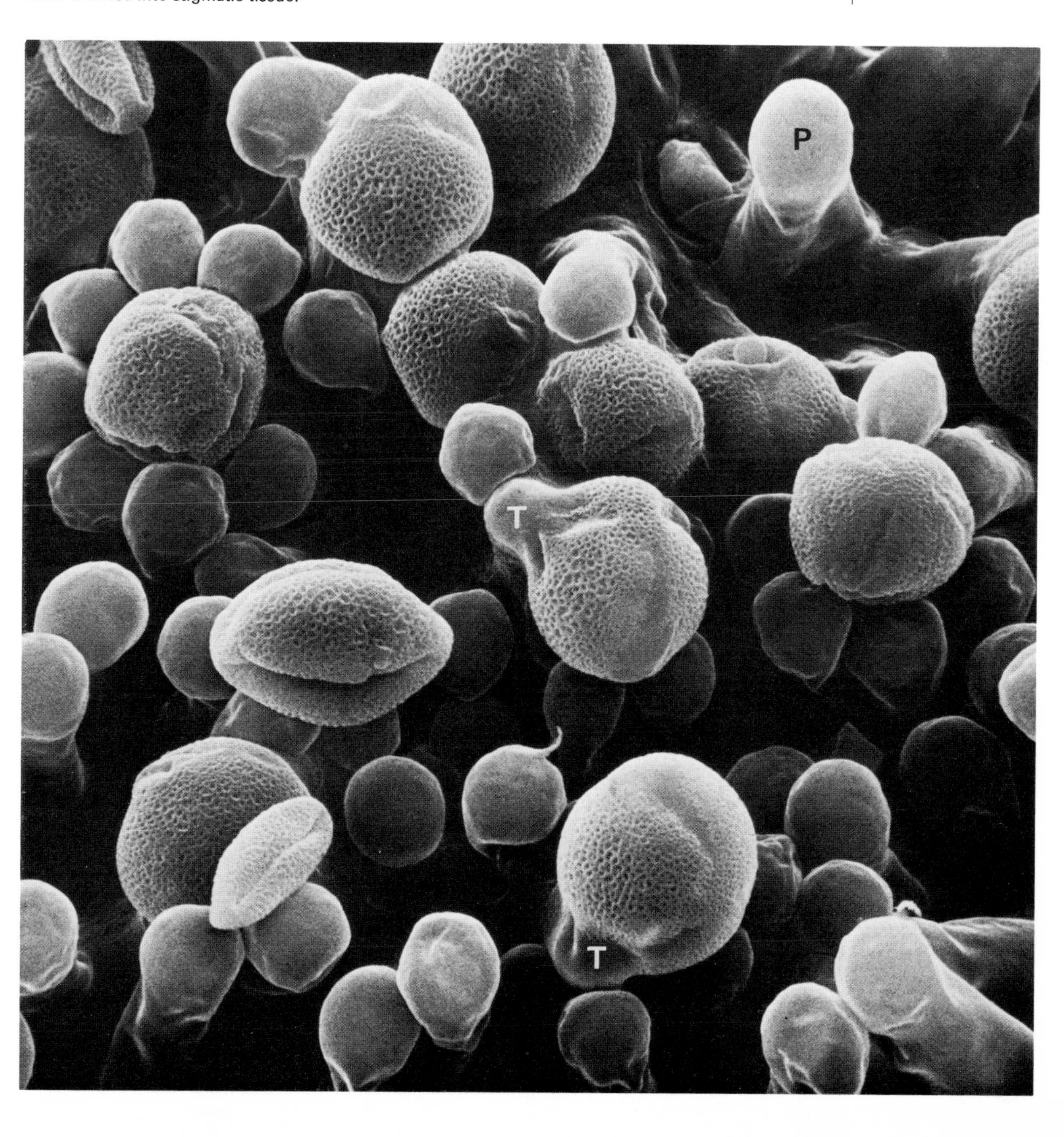

Plate 142
X 4,000

Pin pollen on a thrum stigma in primrose *(Primula)*

The smaller, more oblong pollen grains from pin flowers are shown on a thrum stigma which has shorter stigmatic papillae (**P**) than pin flowers (Plate 141). Note the pollen grain which has germinated to form a pollen tube (**T**).

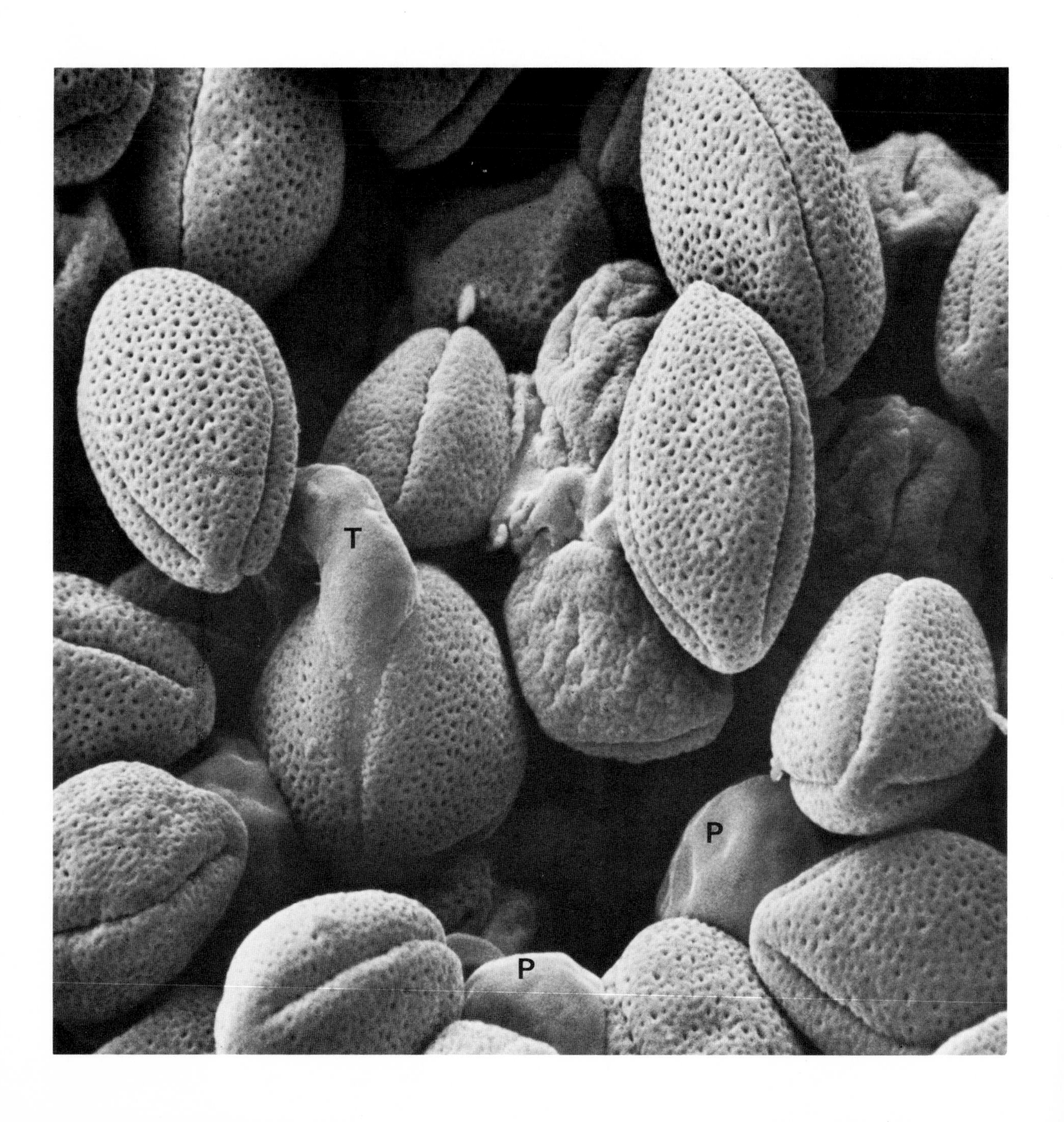

Pollen liberation in *Gomphrena globosa* L.

Plate 143
X 250

This is a side view of the upper part of a stamen (at centre) which has undergone dehiscence. The outer wall of the two pollen sacs, on each side of the stamen, has split longitudinally at a weakened zone of cells, the stomium, along the junction of these sacs. The remains of the internal wall between the pollen sacs can be seen in the photo.

A more magnified view of the pollen is shown in Plate 154.

Plate 144

X 4,200

Pollen grain of *Magnolia campbellii* Hooker

This type of pollen grain has a single elongated deep furrow (colpus) in the wall and illustrates the monocolpate type. The pollen tube breaks through part of this colpus at germination. It is believed that the first angiosperms had this type of pollen (Takhtajan, 1969). Monocolpate pollen occurs in many gymnosperms too.

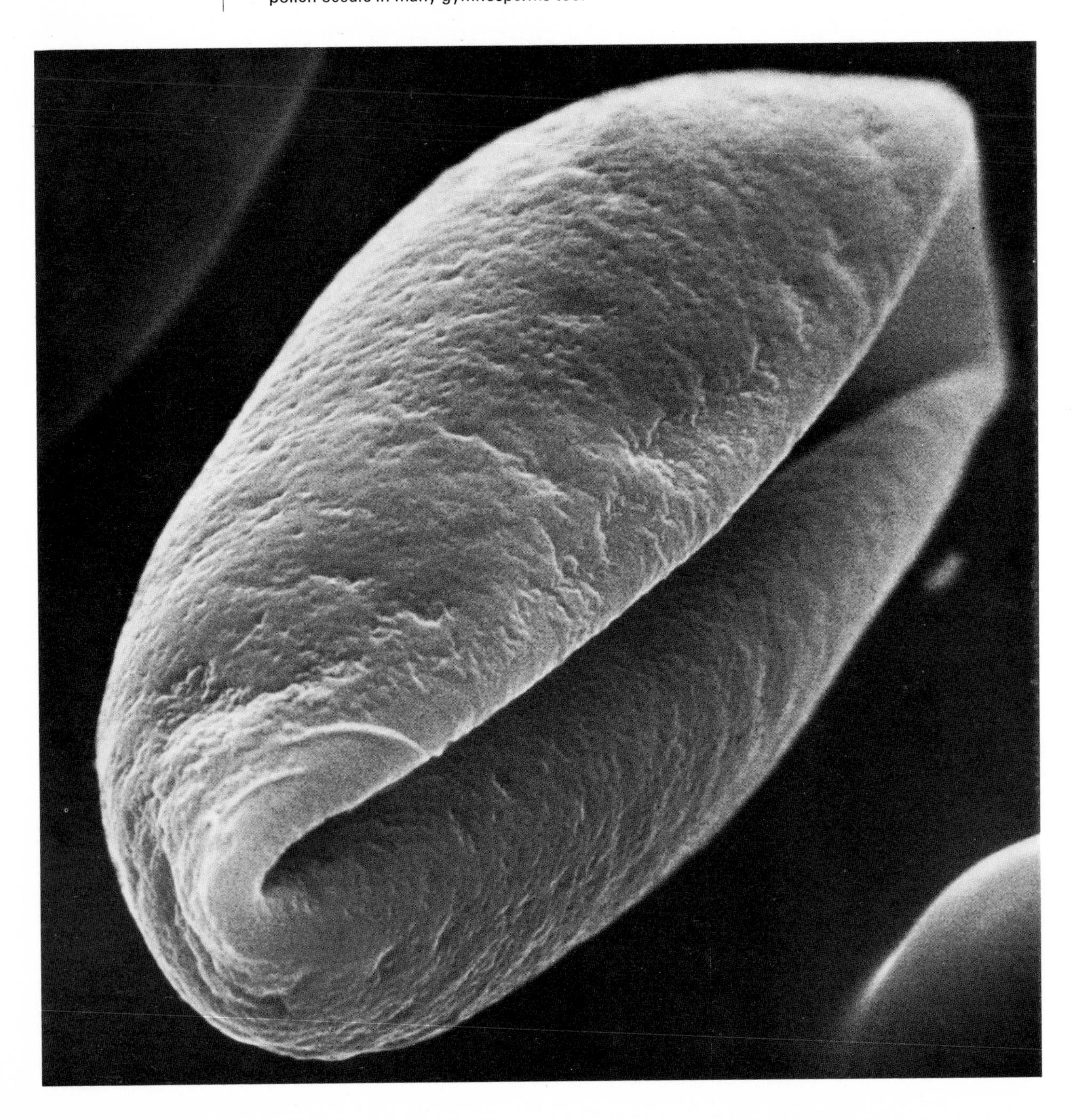

Plate 145 X 3,500

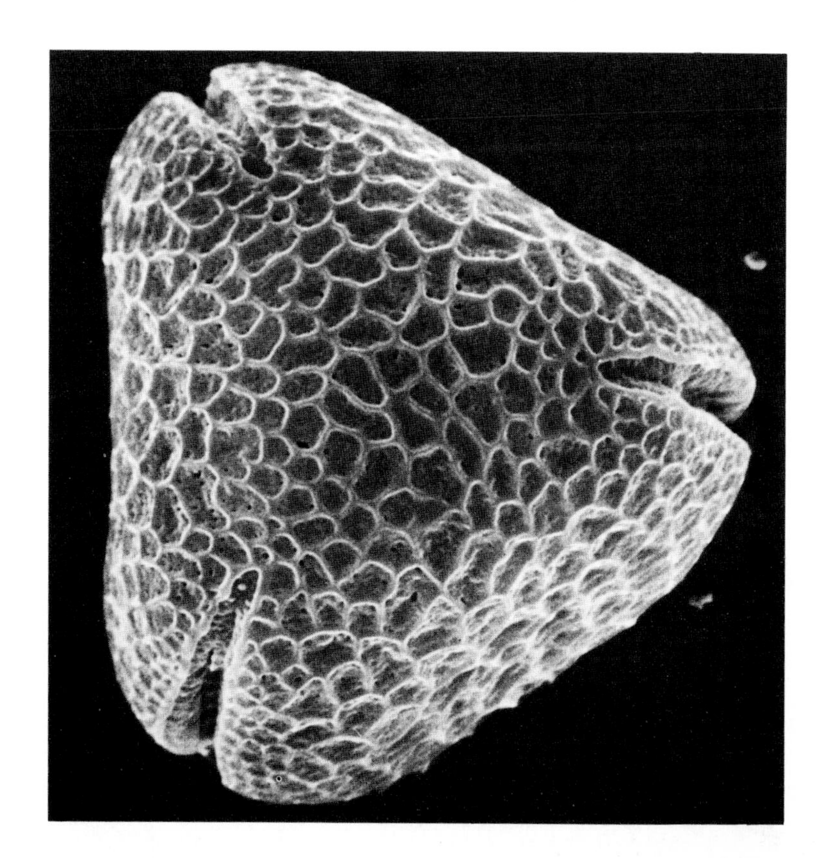

Pollen grain of nasturtium *(Tropaeolum majus)* **in polar and equatorial view**

This plant has tricolpate pollen grains. In this type there are three furrows (colpi) in the wall of the grain, equidistant from one another. The pollen tube germinates through one of these apertures. They are not actual openings but rather regions where the outer wall layer (exine) is very thin. The three furrows are visible in polar view (Plate 145). In equatorial view, the full extent of one of the furrows can be seen (Plate 146). The exine has a reticulate (net-like) surface. There are small, irregularly arranged perforations in the outer layer of the exine, between the ridges of the reticulum.

The tricolpate type of grain is believed to have evolved from the single-furrowed type (monocolpate) which is found in many primitive angiosperms, e.g. *Magnolia* (Plate 144) and *Liriodendron* (Takhtajan, 1969). An advantage of the tricolpate type would seem to be that no matter what side of the grain faced the stigma of the carpel after pollination, a furrow would be relatively close to the stigma. This would permit a more direct entry of the pollen tube into the style.

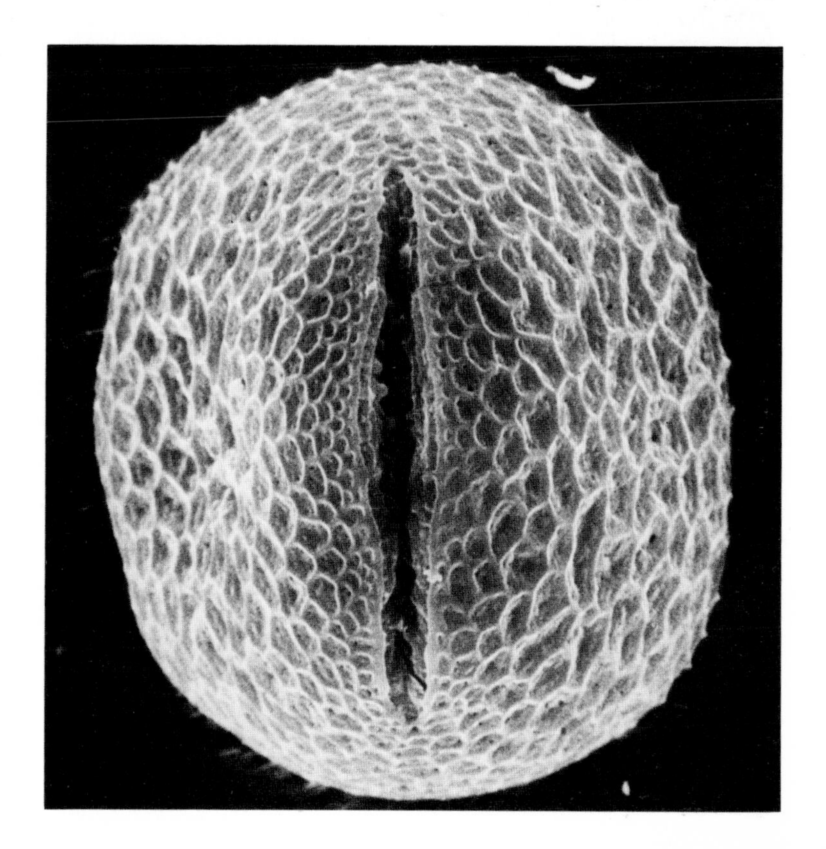

Plate 146 X 3,500

Plate 147
X 3,300

Pollen grain of rose (*Rosa* Tourn. ex L.)

One of the three apertures of this tricolpate grain is visible in this equatorial view. The finely ridged surface has been aptly described as resembling the markings of a fingerprint (Wodehouse, 1935).

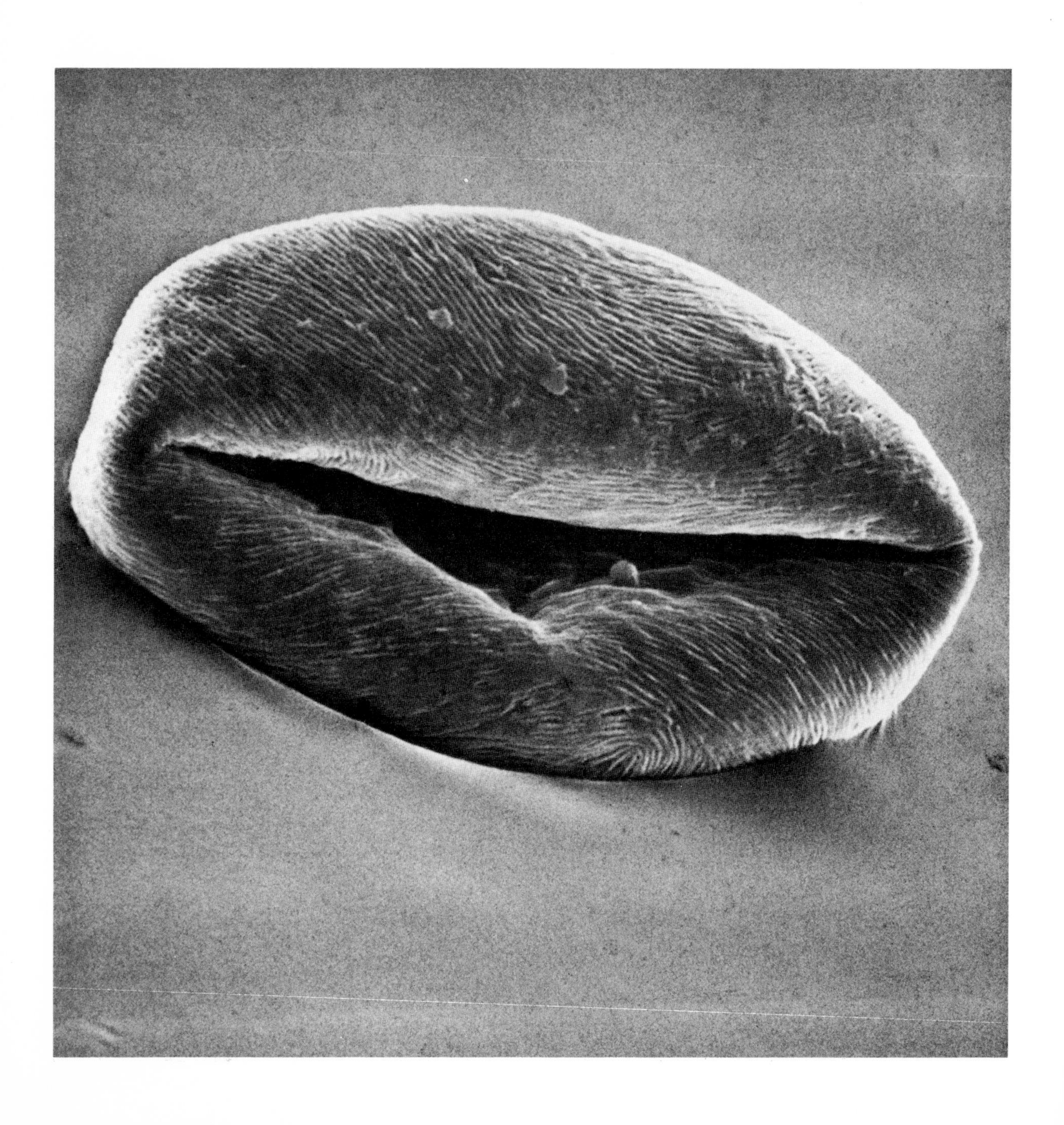

Pollen grain of sowthistle (*Sonchus oleraceus* L.)

Plate 148
X 5,000

This member of the daisy family (Compositae) resembles nasturtium (Plates 145, 146) in having three apertures on each grain. They are a circular shape, rather than a furrow, and such pollen is termed triporate. Two of these pores (upper left and lower right) are in the photo.

The coarse reticulum of ridges bears spines and the outer layer of the exine contains small perforations.

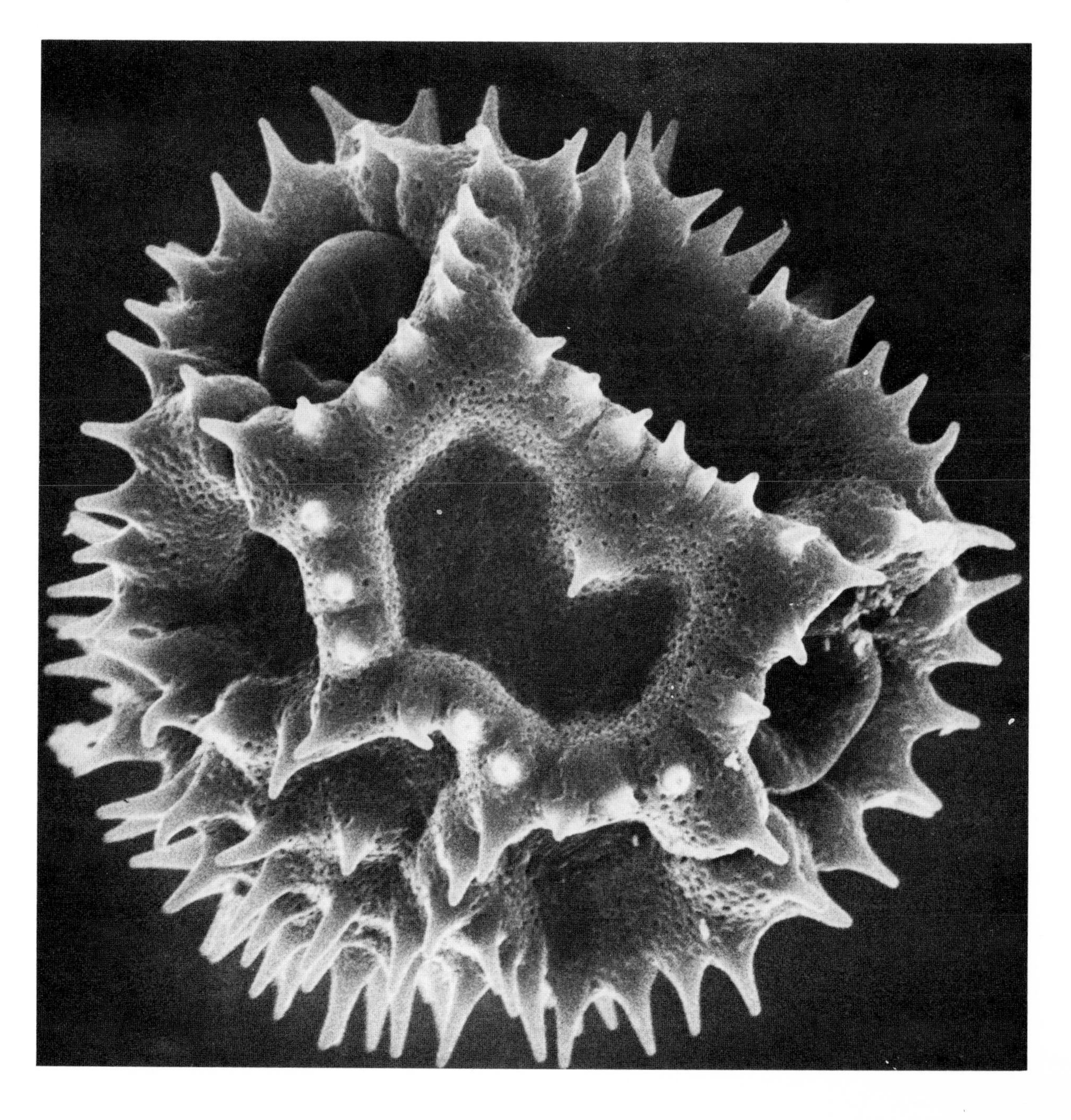

Plate 149
X 3,200

Pollen grain of Kermadec Island pohutukawa (*Metrosideros kermadecensis* W. R. B. Oliver)

These unusual-shaped pollen grains are typical of the family Myrtaceae. In polar view, the grain has a triangular shape, with an aperture at each of the three rounded angles. Each of the three circular apertures lies within the elongated furrow and such pollen is termed tricolporate. The surface of the exine is almost smooth. Changes in the volume of the grains are accomplished by arching or flattening of the polar surfaces. Note the raised "island" (**I**) of wall material at the pole of each grain. This is lacking in some species of *Metrosideros* (McIntyre, 1963).

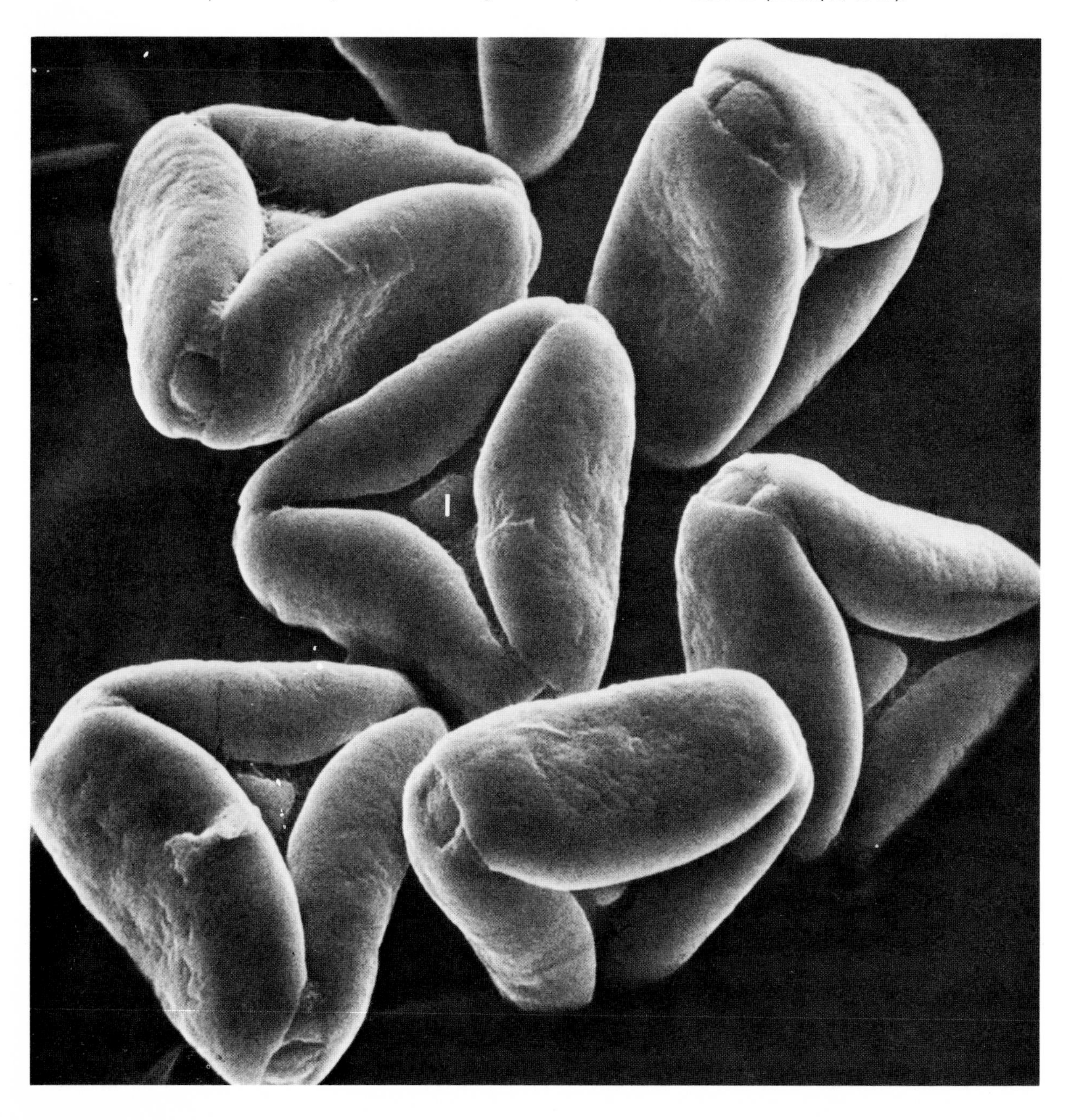

Pollen of storksbill (*Erodium moschatum* (L.) L'Hérit.)

Erodium (family Geraniaceae) has tricolporate pollen. One of the three furrows is visible in the centre of the grain. The reticulum on the surface of the grain forms an attractive interlaced network. This pollen is also illustrated in the frontispiece. There is considerable diversity in pollen morphology within the Geraniaceae.

Plate 150
X 3,800

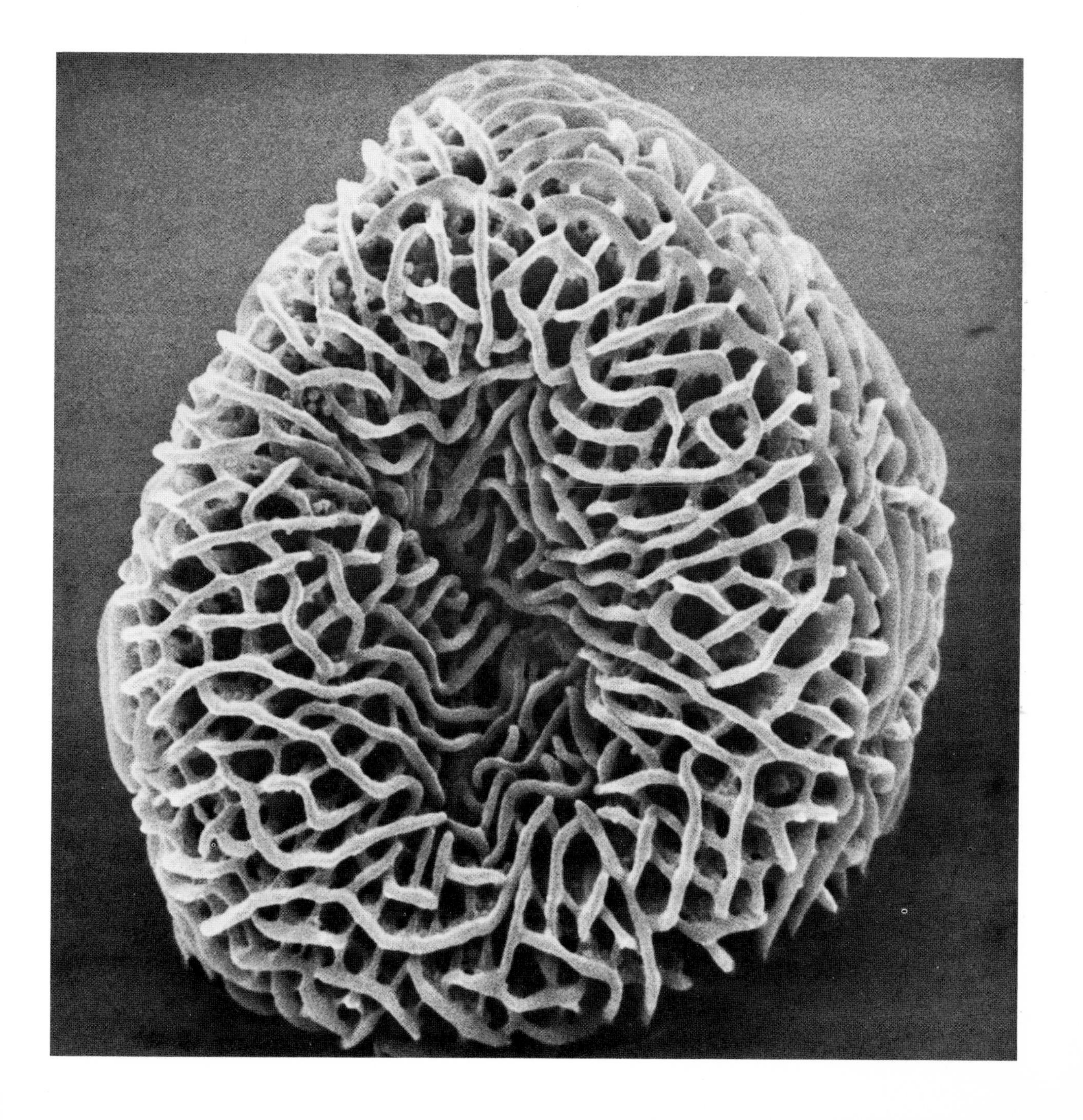

Plate 151
X 5,800

Pollen grain of fennel (*Foeniculum vulgare* Mill.)

Pollen of the carrot family (Umbelliferae) is usually described as tricolporate rather than tricolpate. Although the wall of the grain typically has long furrows, as shown below, the furrow itself encloses a pore in many genera.

The outer layer of the exine of fennel pollen has an intricately woven texture.

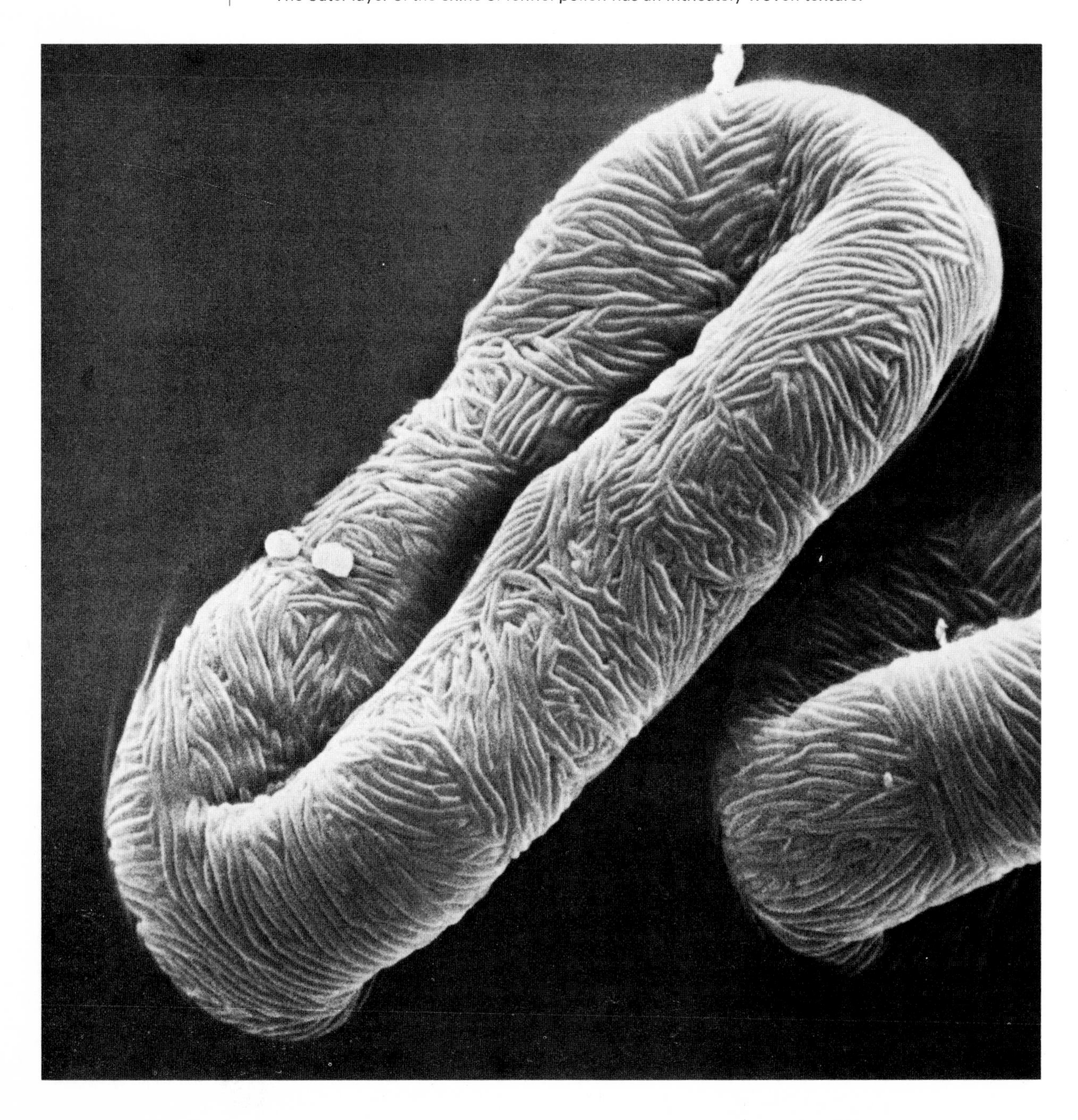

Pollen grain of *Eschscholtzia californica* Cham.

This member of the poppy family (Papaveraceae) has six to seven colpate pollen. The grain is spheroidal and the furrows are long and narrow.

Plate 152
X 4,700

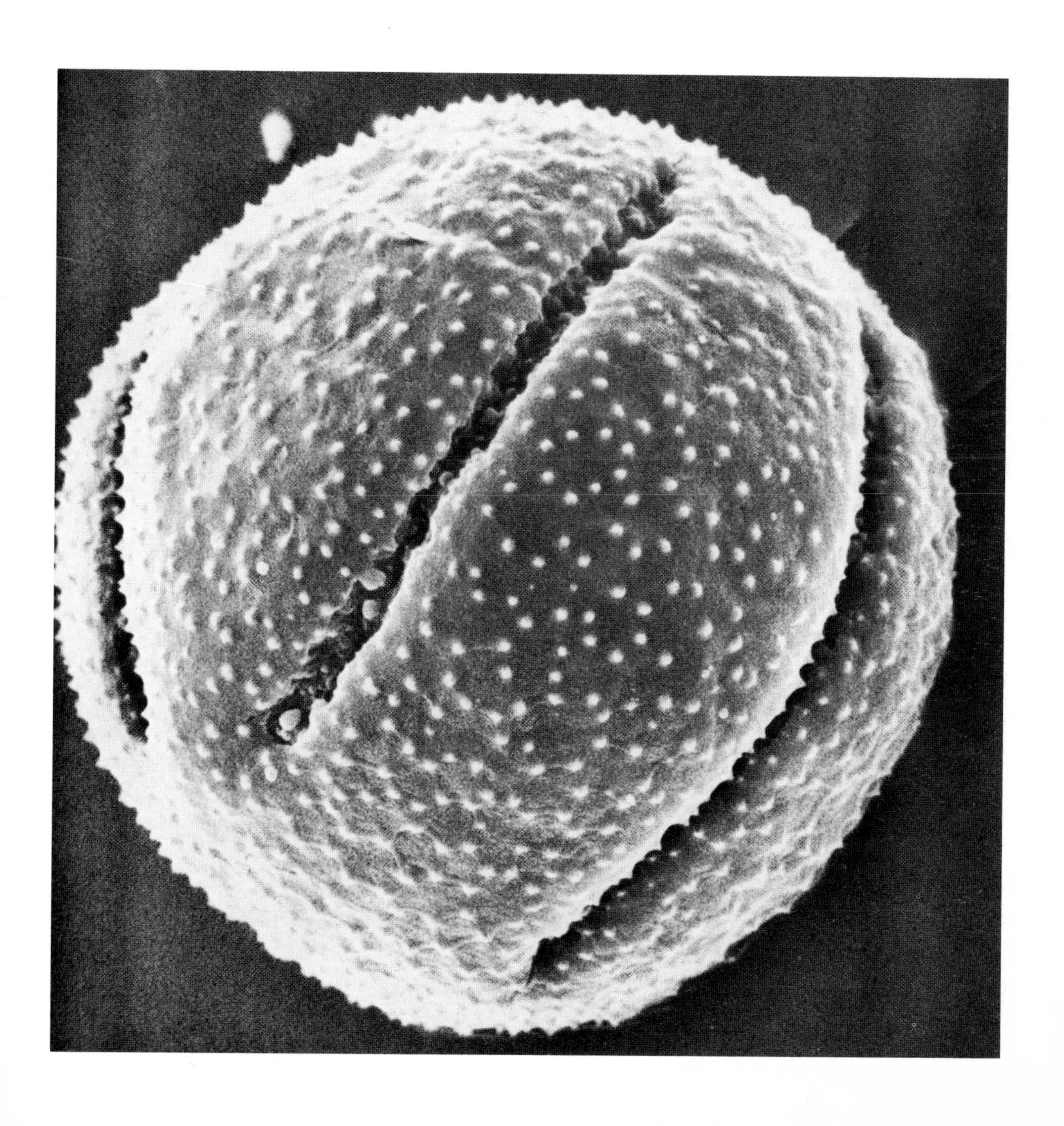

Plate 153

X 6,900

Pollen grain of plantain (*Plantago lanceolata* L.)

The family Plantaginaceae has one of the most advanced types of pollen—the periporate type. In periporate pollen, there are a number of pores (more than three) evenly distributed over the surface of the grain. In this particular species of *Plantago* there is a thickened ring around each pore, known as the annulus (**A**). This species too has a pore which is described as operculate (Faegri and Iversen, 1964) because the so-called aperture has a thick membrane (**M**) which bears the small spines found elsewhere on the surface of the grain and is separated from the rest of the exine by a thin surrounding membrane (**T**).

The number of pores per grain in *Plantago* varies within some species and from species to species from four to more than nine.

There has been some shrinkage of the grain during preparation for SEM examination.

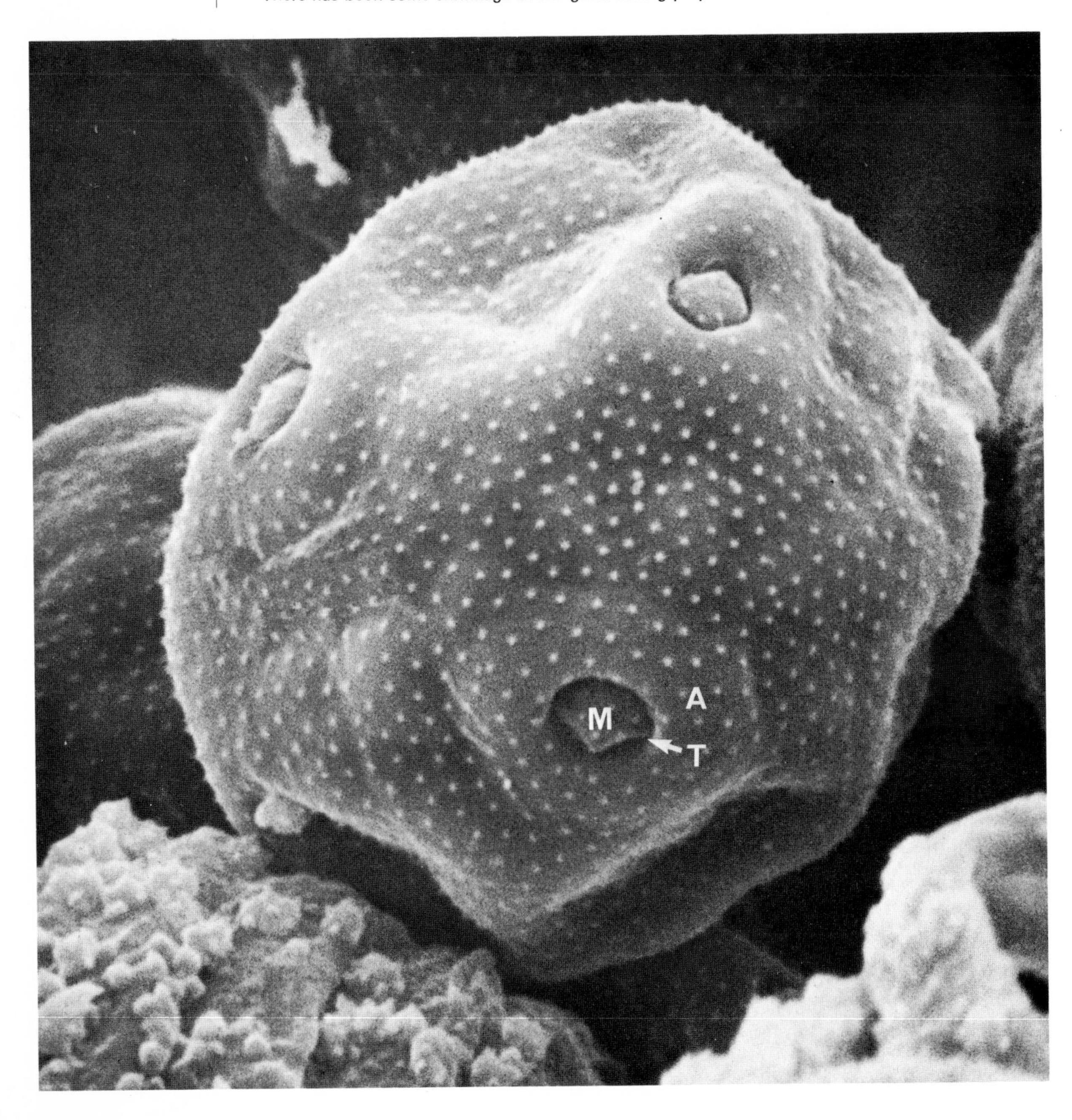

Pollen grains of *Gomphrena globosa* (family Amaranthaceae)

This is a more magnified view of some of the pollen grains shown in Plate 143. Pollen resembles plantain (Plate 153) in being periporate. A large number of pores are recessed between polyhedral ridges which form a coarse reticulum.

Plate 154
X 2,600

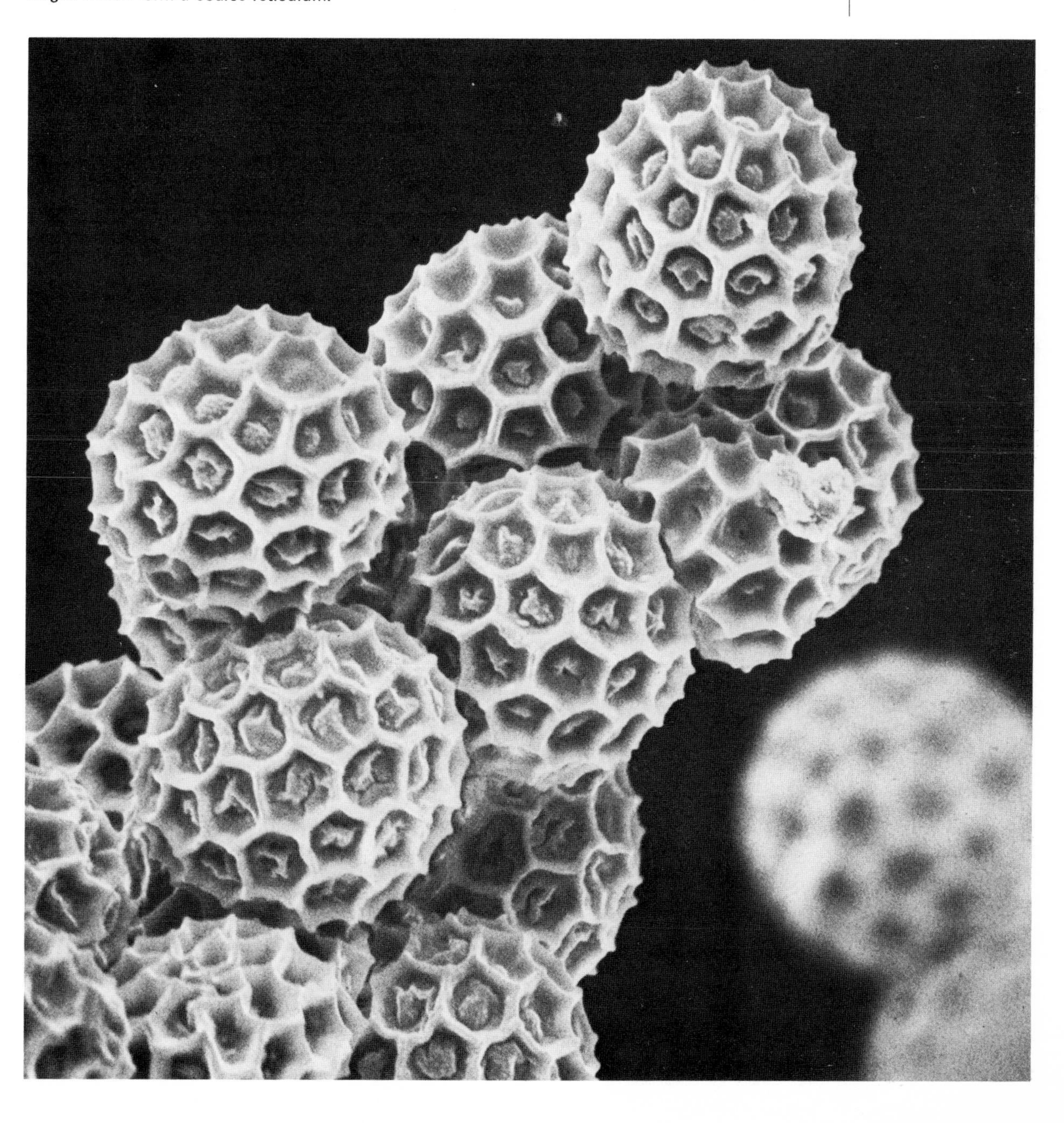

Plate 155

X 3,100

Pollen grain of Canna lily

The monocotyledonous family Cannaceae, which contains the single genus *Canna* L., has inaperturate pollen, in which a furrow or pore is entirely absent. This type of pollen is considered to have evolved from pollen which possessed apertures (Muller, 1970).

The surface of the grain is covered with rounded spines. Among the various varieties of *Canna* there is considerable variation in the ornamentation of the grain from sharp spines to rounded ones to smooth surfaces (Nair, 1970).

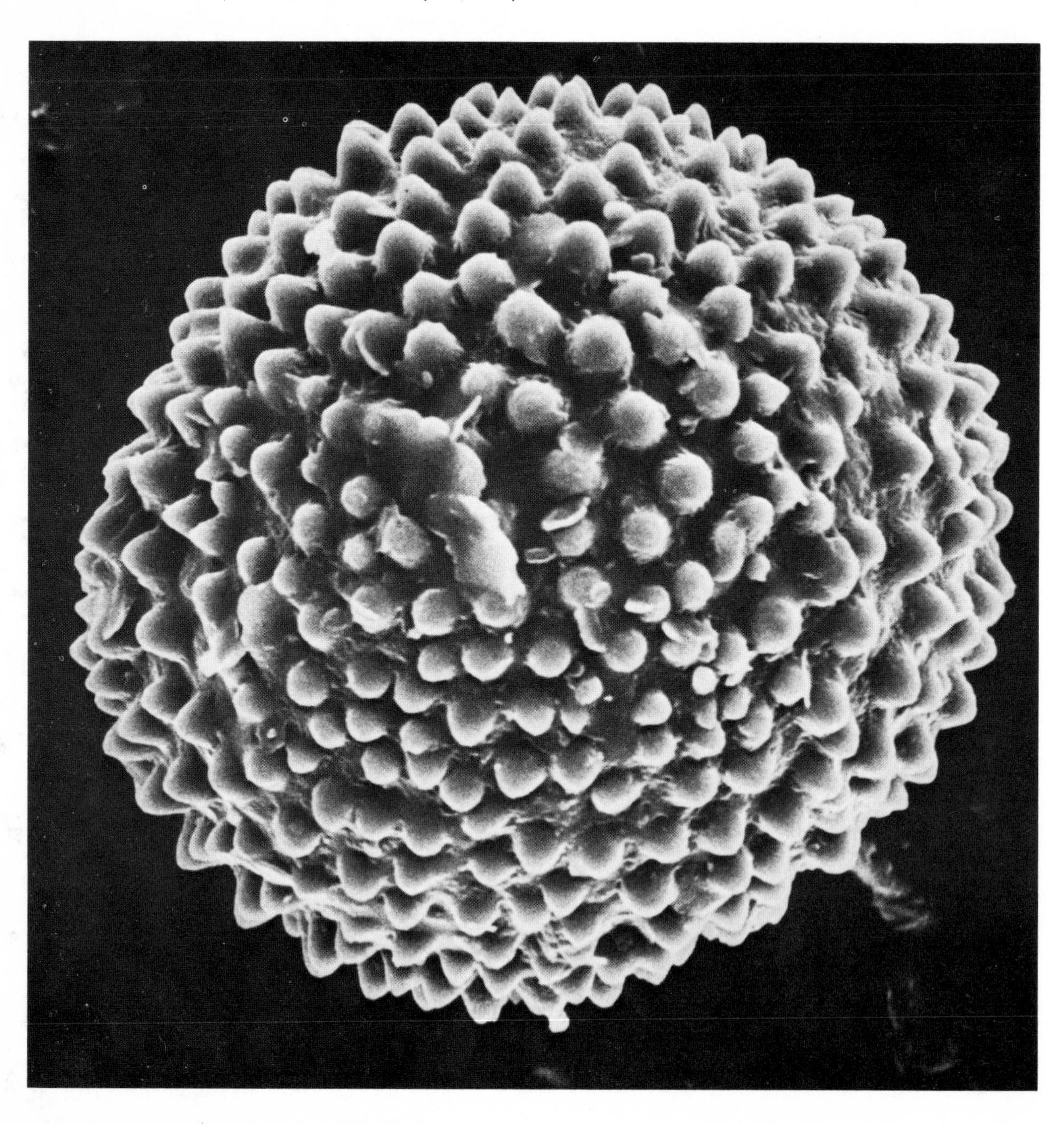

Pollen grain of crabgrass *(Digitaria sanguinalis)*

Plate 156
X 3,000

Pollen of the grass family (Gramineae) is remarkably uniform. The SEM is useful in showing subtle differences between grasses. All grains have a single pore (monoporate type). In crabgrass the circular pore is surrounded by an annulus and is operculate, as in plantain (Plate 153). Grass pollen has a distinctive slightly rough and stippled texture (Wodehouse, 1935).

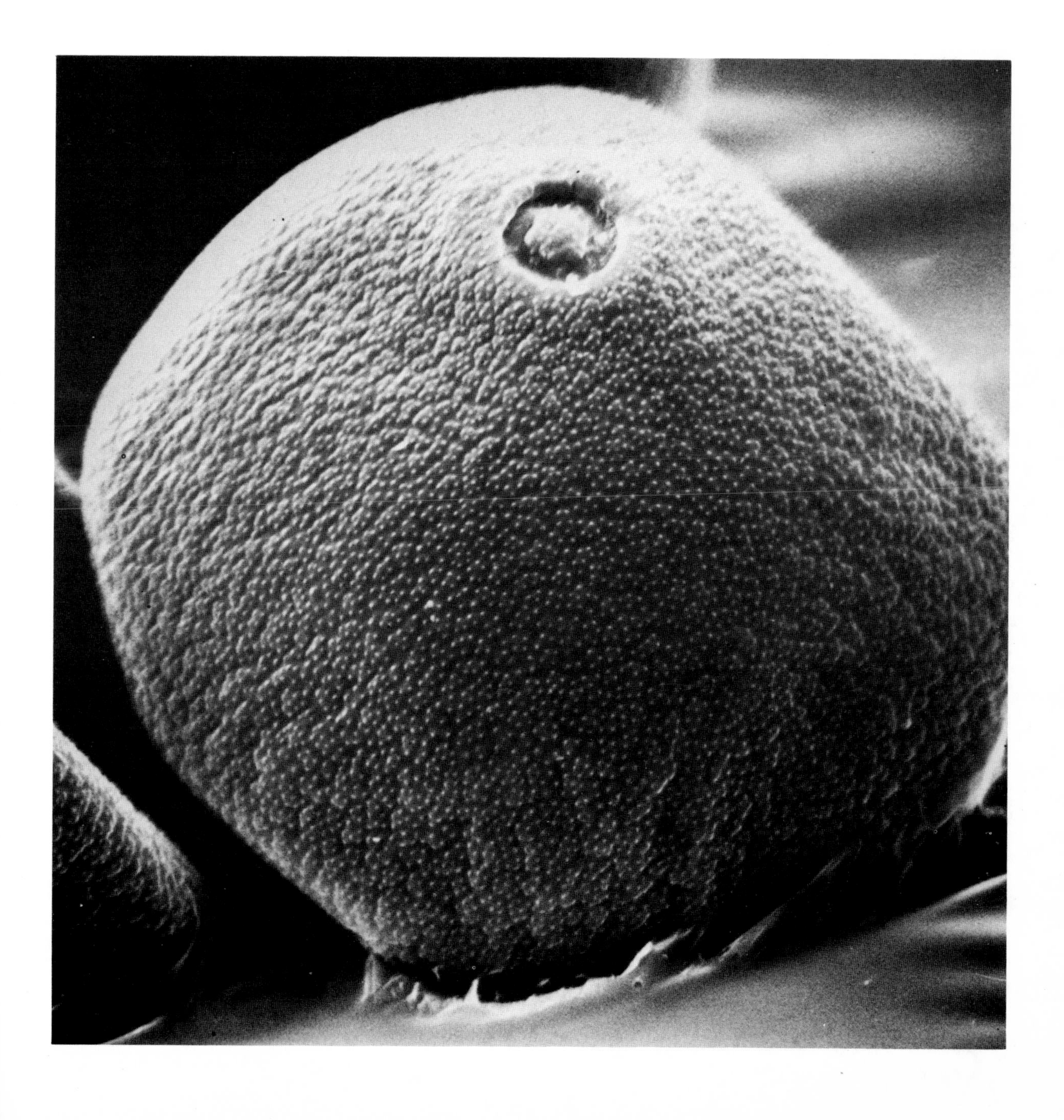

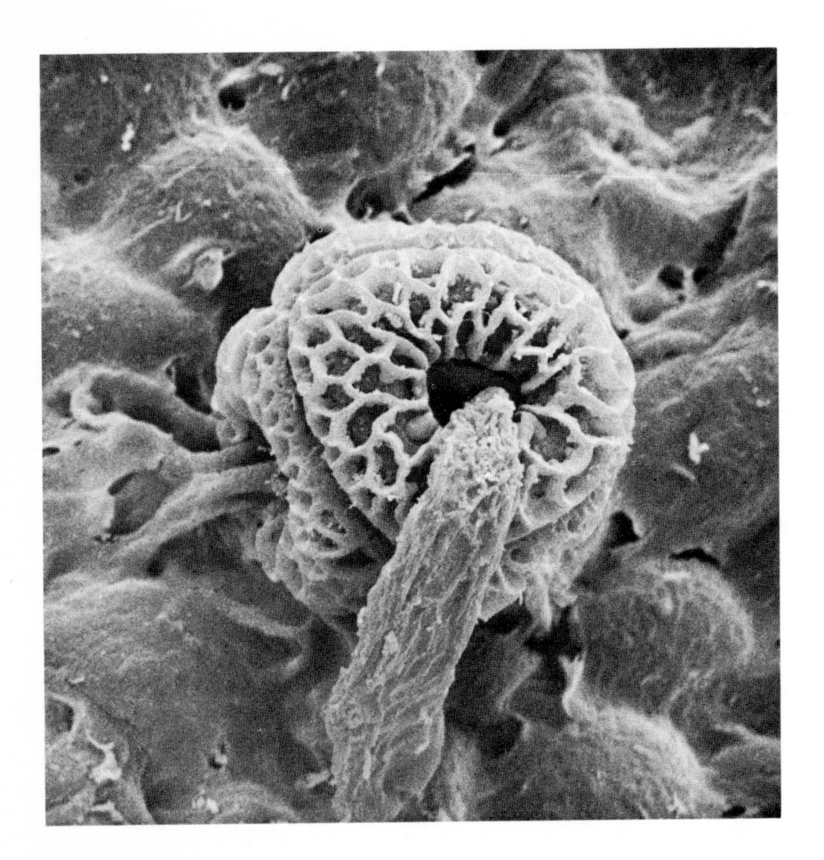

Plate 157

X 1,550

Pollen on the stigma of
Pseudowintera axillaris
(J. R. et G. Forst.) Dandy

In this dicotyledonous plant and other members of the primitive vesselless family Winteraceae, pollen grains remain in groups of four. In most angiosperms, the four pollen grains that are formed by the meiotic division of a pollen mother cell, separate from one another at an early stage in their development. However, permanent pollen tetrads occur in a number of unrelated families. Pollen tubes at centre and left extend from the germ pores of the two pollen grains on the near side of the tetrad. Each pollen grain of *Pseudowintera* has a single germ pore situated in the centre of its external wall. The pollen tubes grow down through the upper part of the carpel until they reach the ovules (Plate 159).

Plate 158 X 360

Ovule of *Pseudowintera*

A carpel has been cut open to reveal one of the ovules. The ovule is attached by its stalk (funicle) to a part of the internal wall of the carpel, termed the placenta. The funicle is hidden by part of the carpel wall at the top of the photo. The central nucellus tissue of the ovule is enclosed by two coverings, the inner and outer integuments. In this plant the inner integument extends beyond the outer. The tip of the inner integument has a flattened tubular opening, the micropyle (**M**), through which the pollen tube passes (Plate 159). The integuments of angiosperms do not have the finger-like extensions which occur on the integument of *Pinus* (Plate 93).

Most angiosperm ovules have two integuments; some have a single integument. There is evidence that two integuments represent the primitive condition and that the single integument has arisen by fusion of two separate integument primordia in some plants and by evolutionary loss of one of the integuments in others.

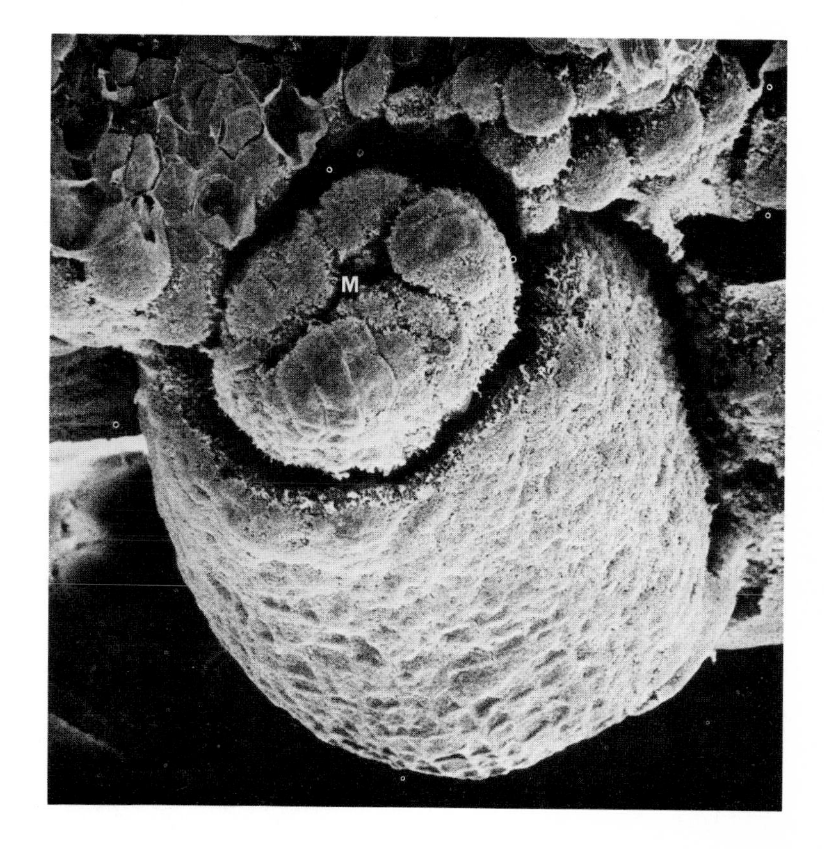

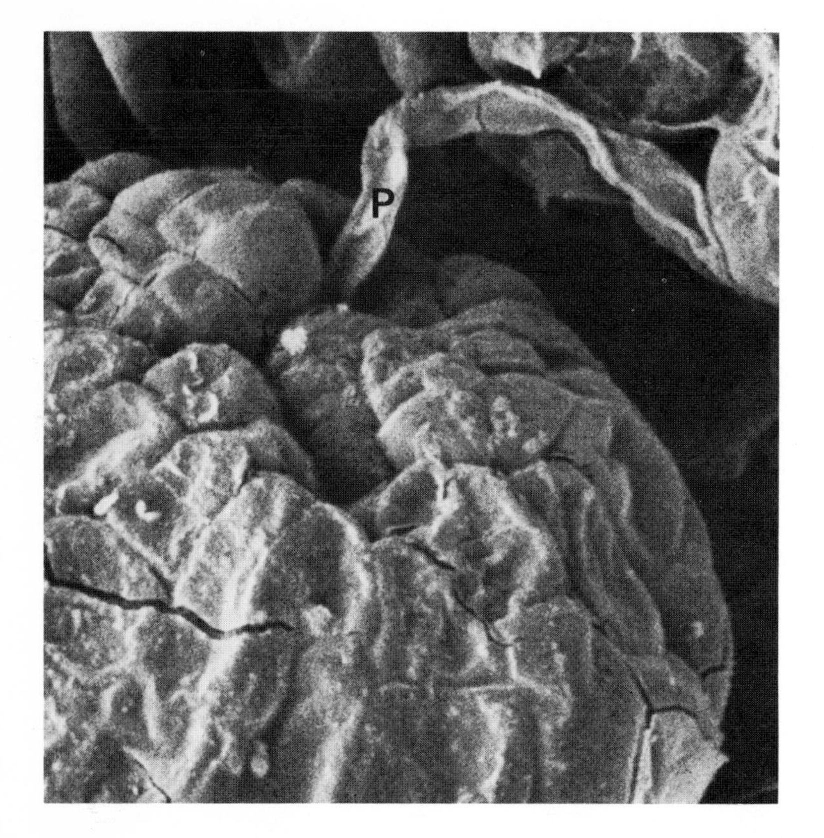

Plate 159

X 1,400

Pollen tube extending down the micropyle in *Pseudowintera*

In *Pseudowintera* the micropyle of the ovule faces upwards. The pollen tube (**P**) has grown through the upper part of the carpel to reach the ovule and has entered it via the micropyle. It grows down the micropyle and reaches the nucellus and passes through nucellar tissue until it reaches the female gametophyte tissue (embryo sac). The two male gametes are then discharged from near the tip of the pollen tube and double fertilisation takes place. One gamete fuses with the female gamete (ovum) to form the young one-celled embryo (zygote) and the other fuses with the central fusion nucleus in the embryo sac. As a result of this second fusion, endosperm tissue is formed which provides nutriment for the embryo. After fertilisation the ovule is known as the seed.

Young embryo from the seed of shepherd's purse (*Capsella bursa-pastoris* Medic.)

Plate 160

X 800

The embryo of this dicotyledonous plant of the mustard family (Cruciferae) is at the torpedo-shaped stage of development. In extraction of the embryo from the seed the suspensor has been lost. The suspensor is a string of six to ten cells which are attached to the tip of the radicle (embryo root) at the top of the photo. The suspensor pushes the growing embryo deeper into the nutritive endosperm tissue. It also functions in absorbing and transporting nutrients to the embryo from the endosperm and surrounding nucellar tissue (Cutter, 1971). At a later stage the suspensor cells become crushed by the growth of the embryo, with the exception of the terminal cell, which divides to form part of the root apex and root cap of the embryo. The two cotyledons (seed leaves) enclose the shoot apex, from which the foliage leaves are derived.

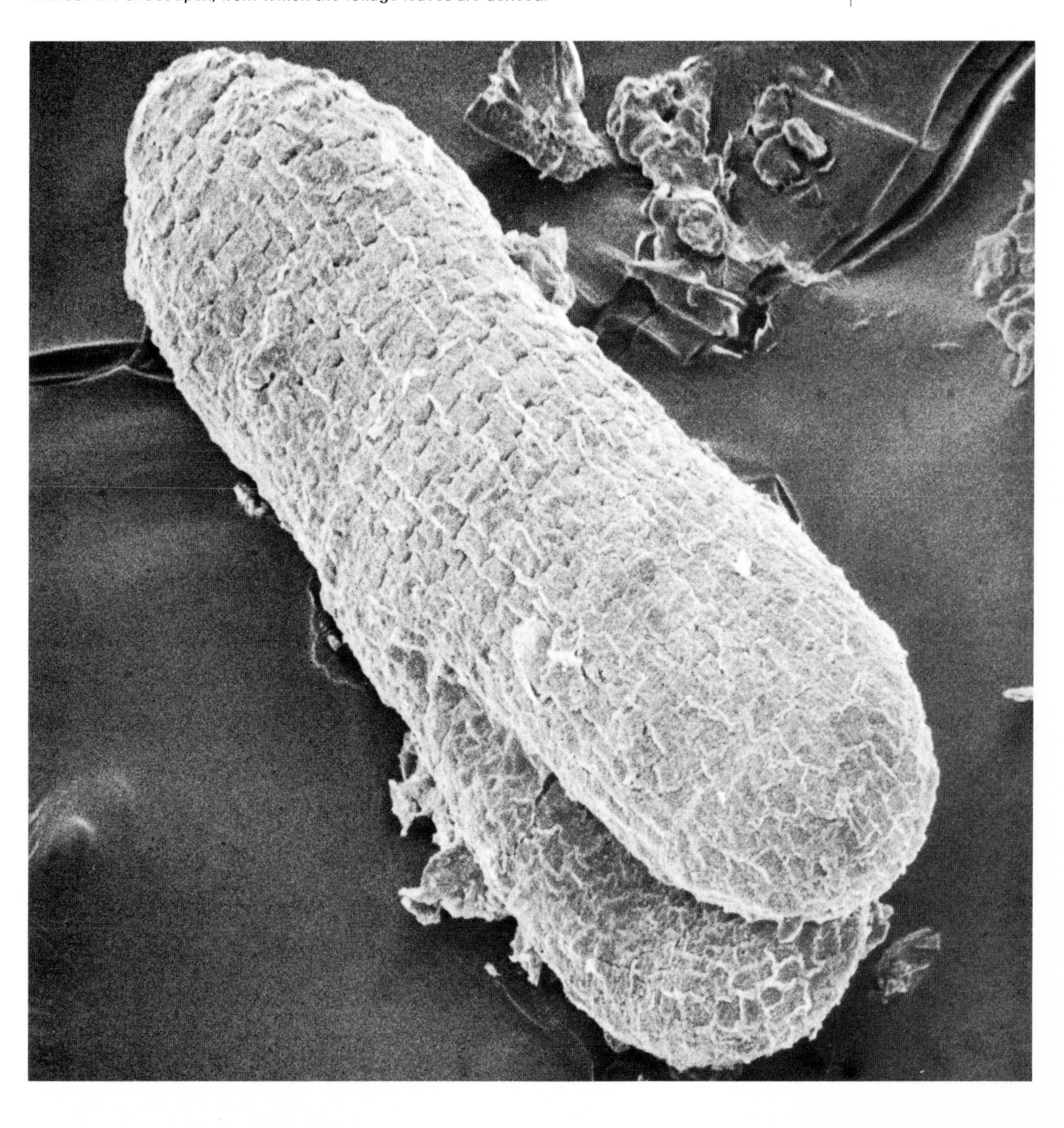

Plate 161
X 240

Older embryo from the seed of shepherd's purse *(Capsella bursa-pastoris)*

At this stage the embryo is nearly mature. In *Capsella* the embryo sac, containing the nutritive endosperm tissue, is a curved U-shape. To accommodate to this shape, the embryo has become folded in half during its later growth. The shoot apex, from which foliage leaf primordia have not yet been initiated, is still enclosed by the cotyledons. Comparison with the earlier stage (Plate 160) reveals that there has been considerable expansion of the hypocotyl region between the base of the cotyledons and the root. In the centre of the embryo, between the root and shoot apices, there would be well-developed procambial tissue from which the primary vascular tissue is derived.

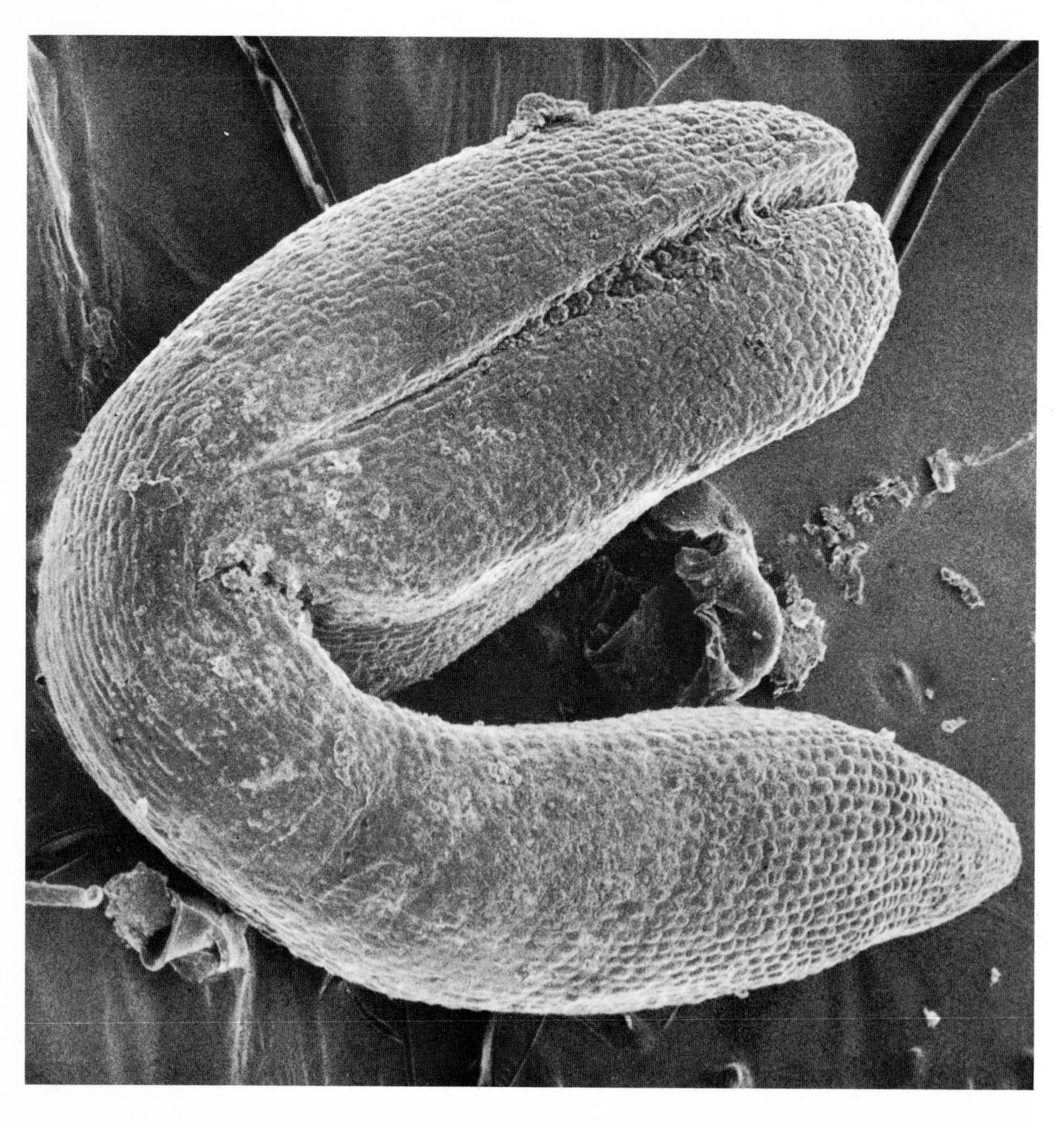

Fruit of groundsel (*Senecio vulgaris* L.)

Plate 162
X 70

Fruits of the daisy family (Compositae) are of the achene type. An achene is a dry, one-seeded fruit which has no special method of opening to liberate the seed. At the top of the fruit in Plate 162 is a pappus of hairs. These give buoyancy to this wind-dispersed fruit. The pappus is confined to the Compositae and is equivalent to the calyx (sepals) of other flowers. A fruit is the ripened ovary of a flower. As the ovary is an inferior one in *Senecio*, the pappus is at the morphologically upper part of the fruit.

Plate 163 is a more magnified view of part of the fruit surface.

Plate 163
X 300

More magnified view of fruit surface of groundsel

See Plate 162.

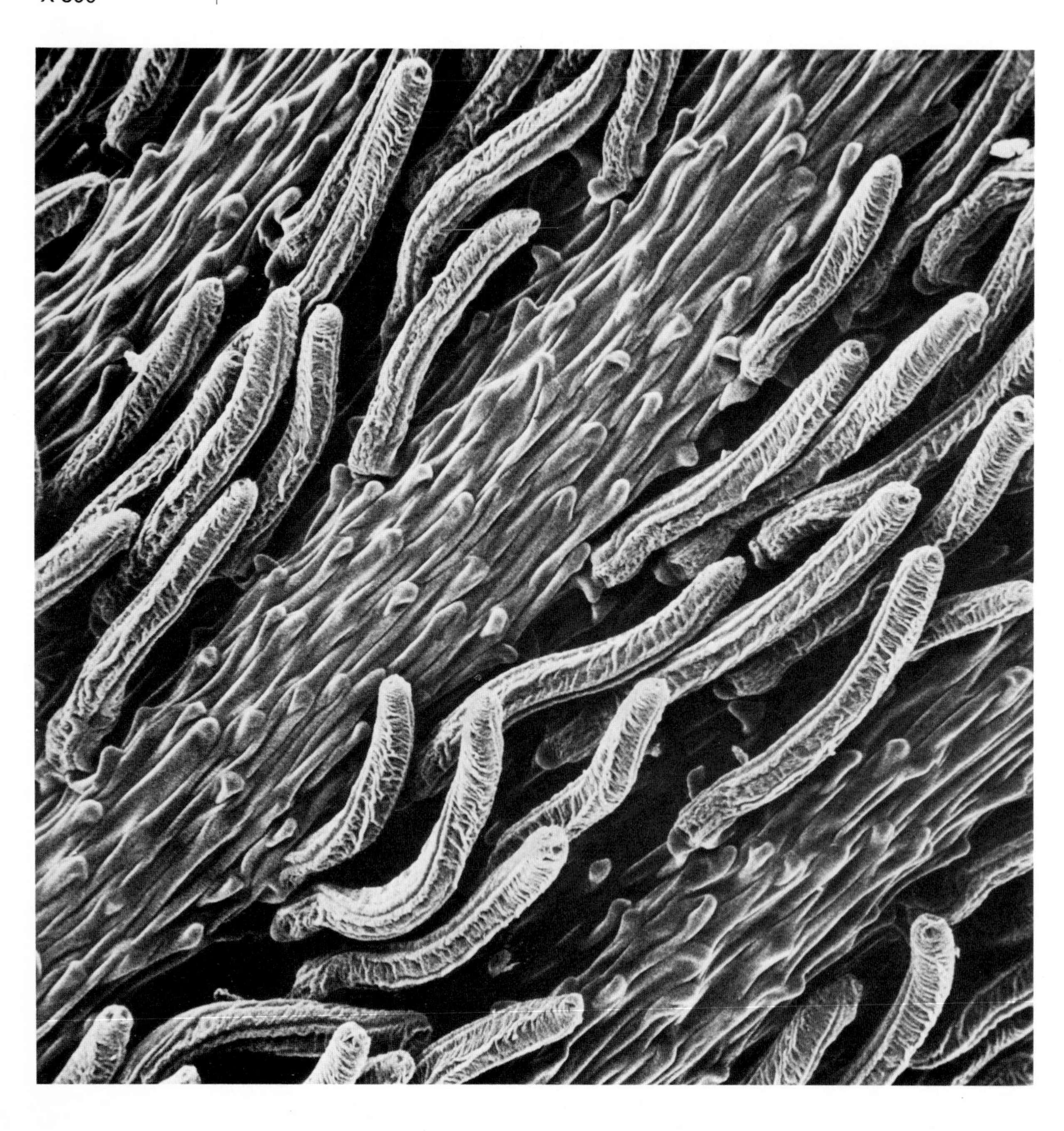

References

Alexopoulos, C. J., *Introductory mycology*, 2nd edition, John Wiley and Sons, New York, 1962.

Bierhorst, D. W., *Morphology of vascular plants*, The Macmillan Company, New York, 1971.

Campbell, E. O., "*Marchantia* species of New Zealand", *Tuatara*, 1965, **13**, pp. 122-36.

Cutter, E. G., *Plant anatomy: experiment and interpretation*, Part 1, *Cells and tissues*, Edward Arnold, London, 1969.

Cutter, E. G., *Plant anatomy: experiment and interpretation*, Part 2, *Organs*, Edward Arnold, London, 1971.

Eames, A. J., *Morphology of the angiosperms*, McGraw-Hill Book Company, New York, 1961.

Eames, A. J. and MacDaniels, L. H., *Introduction to plant anatomy*, 2nd edition, McGraw-Hill Book Company, New York, 1947.

Esau, K., *Plant anatomy*, 2nd edition, John Wiley and Sons, New York, 1965.

Esau, K., "The phloem", *Encyclopedia of plant anatomy*, Volume 5, Part 2, Gebruder Borntraeger, Berlin, 1969.

Faegri, K. and Iversen, J., *Textbook of pollen analysis*, 2nd edition, Blackwell, Oxford, 1964.

Harris, W. F., "A manual of the spores of New Zealand pteridophyta", *Bull.N.Z.Dept.scient.ind.Res.*, 1955, **116**.

Hatch, M. D. and Slack, C. R., "Photosynthetic CO_2-fixation pathways", *A.Rev.Pl.Physiol.*, 1970, **21**, pp. 141-62.

Ingold, C. T., *Spore liberation*, Oxford University Press, London, 1965.

Kaufman, P. B., Bigelow, W. C., Schmid, R. and Ghosheh, N. S., "Electron microprobe analysis of silica in epidermal cells of *Equisetum*", *Am.J.Bot.*, 1971, **58**, pp. 309-16.

Ledbetter, M. C. and Porter, K. R., *Introduction to the fine structure of plant cells*, Springer-Verlag, Berlin, 1970.

Lewin, J. and Reimann, B. E. F., "Silicon and plant growth", *A.Rev.Pl.Physiol.*, 1969, **20**, pp. 289-304.

MacRobbie, E. A. C., "Phloem translocation. Facts and mechanisms: a comparative survey", *Biol.Rev.*, 1971, **46**, pp. 429-81.

Martin, W. and Child, J., *New Zealand lichens*, A. H. and A. W. Reed, Wellington, New Zealand, 1972.

McIntyre, D. J., "Pollen morphology of New Zealand species of Myrtaceae", *Trans.R.Soc.N.Z.Bot.*, 1963, **2**, pp. 83-107.

Metcalfe, C. R. and Chalk, L., *Anatomy of the dicotyledons*, Volumes 1 and 2, Oxford University Press, London, 1950.

Meylan, B. A. and Butterfield, B. G., "Three-dimensional structure of wood. A scanning electron microscope study", A. H. and A. W. Reed, Wellington, New Zealand (Chapman and Hall, London; Syracuse University Press, New York), 1972.

Monroe, J. H., "Light- and electron-microscopic observations on spore germination in *Funaria hygrometrica*", *Bot.Gaz.,* 1968, **129**, pp. 247-58.
Mozingo, H. N., Klein, P., Zeevi, Y. and Lewis, E. R., "Scanning electron microscope studies on *Sphagnum imbricatum*", *Bryologist,* 1969, **72**, pp. 484-8.
Muller, J., "Palynological evidence on early differentiation of angiosperms", *Biol.Rev.,* 1970, **45**, pp. 417-50.
Nair, P. K. K., *Pollen morphology of angiosperms,* Vikas Publications, Delhi, 1970.
Nicholson, N. L. and Briggs, W. R., "Translocation of photosynthate in the brown alga *Nereocystis*", *Am.J.Bot.,* 1972, **59**, pp. 97-106.
Pant, D. D. and Khare, P. K., "Epidermal structure of Psilotales and stomatal ontogeny of *Tmesipteris tannensis* Bernh.", *Ann.Bot.,* 1971, **35**, pp. 151-7.
Parihar, N. S., *An introduction to embryophyta,* Volume 1, *Bryophyta,* 5th edition, Central Book Depot, Allahabad, India, 1965.
Proskauer, J., "On the peristome of *Funaria hygrometrica*", *Am.J.Bot.,* 1958, **45**, pp. 560-3.
Raven, P. H. and Curtis, H., *Biology of plants,* Worth Publishers, New York, 1970.
Smith, G. M., *Cryptogamic botany,* Volume 1, 2nd edition, McGraw-Hill Book Company, New York, 1955.
Stoermer, E. F., Pankratz, H. S. and Bowen, C. C., "Fine structure of the diatom *Amphipleura pellucida.* II. Cytoplasmic fine structure and frustule formation", *Am.J.Bot.,* 1965, **52**, pp. 1067-78.
Stone, I. G., "The gametophyte of the Victorian Blechnaceae I, *Blechnum nudum* (labill.) Luerss.", *Aust.J.Bot.,* 1961, **9**, pp. 20-36.
Stone, I. G., "The gametophyte of the Victorian Blechnaceae II, *Doodia aspera* R.Br., *D.media* R.Br., and *D.caudata* R.Br.: a comparison with three extra-Australian genera, *Brainea, Sadleria,* and *Woodwardia*", *Aust.J.Bot.,* 1969, **17**, pp. 31-57.
Takhtajan, A., *Flowering plants. Origin and dispersal,* Oliver and Boyd, Edinburgh, 1969.
Thurston, E. L. and Lersten, N. R., "The morphology and toxicology of plant stinging hairs", *Bot.Rev.,* 1969, **35**, pp. 393-412.
Troughton, J. and Donaldson, L. A., *Probing plant structure,* A. H. and A. W. Reed, Wellington, New Zealand (Chapman and Hall, London; McGraw-Hill Book Company, New York), 1972.
Watson, E. V., "Famous plants, 6. *Funaria*", *New Biology,* 1957, **22**, pp. 104-24.
——, *The structure and life of bryophytes,* 3rd edition, Hutchinson and Company, London, 1971.
Weier, T. W., Stocking, C. R. and Barbour, M. G., *Botany: an introduction to plant biology,* 4th edition, John Wiley and Sons, New York, 1970.
Wodehouse, R. P., *Pollen grains,* McGraw-Hill Book Company, New York, 1935.
Zwanneveld, C. H., *A freeze-drier for biological specimens,* PEL Report, New Zealand Department of Scientific and Industrial Research, 1973 (in preparation).

Index

(*Please note that these numbers refer to pages rather than plates.*)